The Mysterious Number 6174
One of 30 Amazing Mathematical Topics in Daily Life

By

Yutaka Nishiyama

Gendai Sugakusha, Co.

Osaka University of Economics Research Series, Vol. 79

The Mysterious Number 6174: One of 30 Amazing Mathematical Topics in Daily Life

Osaka University of Economics Research Series, Vol. 79

Copyright © 2013 Yutaka Nishiyama.

All rights reserved.

No part of this book may be reproduced, stored in a retrieval system, or transmitted in any form or by any means, without prior written permission from the publisher, except in the case of brief quotations.

First published: July 2013

Published in Japan by Gendai Sugakusha, Co.

Nishiteranomae, Shishigatani, Sakyo, Kyoto 606-8425, Japan

http://www.gensu.co.jp/

ISBN 978-4-7687-6174-8

Translated by ThinkSCIENCE, Inc.

Cover design by Takako

Preface

This book is designed to reach a wide audience. Not only is it written for readers of all ages, from young to old, but it is also targeted to those who have a dislike of maths. The aim is to show just how wonderful maths really is! This book is a compilation of 30 articles that I originally issued as a series entitled "Enjoying Maths" for the Japanese magazine *Rikeieno Sugaku* (*Mathematics for Science*). All articles are independent of each other and readers can move throughout the book as they please, exploring the wonder of maths as they go.

The book contains my original articles as well as references to other unique articles about the use of maths in daily life. In terms of my original articles, learn about boomerangs, one of my life works. Find out why a boomerang comes back to you and follow the instructions on how to make and throw paper boomerangs to explore this practically. In fact, people all over the world have already tried this: the original article has been translated into 70 languages so far, as can be seen on one of my websites at http://www.kbn3.com/bip/index2.html.

Also find out why so many flowers are five-petalled. Why do we have five digits on our hands and feet? I will explain my hypothesis about the mysterious occurrences of the number "5" hidden in nature. I also explain the development of a theorem about constructing fixed points and the random-dot pattern that facilitated the invention of this new theorem.

This book also contains unique articles about the use of maths in daily life. Find out the answers to questions such as how can two separate light switches in the hall of our house as well as on the upstairs landing work to turn on and off the same light? Why do some electric fans appear to rotate backwards? Why are eggs oval shaped? I think about these issues mathematically, and try to pose intriguing questions you won't find in any textbooks and provide understandable answers. Although the maths community often discusses how to solve problems more efficiently, I like to emphasize the enjoyment and the importance of discovering new mathematics problems in daily life.

Many puzzles dealing with diagrams or numbers are explained. One example is the block overhang problem: is it possible to stagger building blocks by more than the width of one block? Other examples include why a Möbius strip produces one large loop when cut instead of separating into two pieces. Learn about the new face of hexaflexagons and how they work. Do you know about Miura folding, where a single movement can be used to open and close a sheet of paper? How about the increasing and decreasing of areas in card magic? Do you know how it's done?

Then try some really fun uses of numbers that will impress your friends. Take any four digits, reorder them into the largest and smallest numbers possible, take the difference between the largest and the smallest, and repeat until the sequence eventually arrives at number 6174. A summary of this article appears on the University of Cambridge website, at http://plus.maths.org/issue38/features/nishiyama/.

Learn about why maths is really useful in everyday life and how it is applied. For example, the planimeter measures areas just by tracing a closed curve. So, maths is not just a way some people wile away time. It really does have lots of practical applications. Do you know the inside of random numbers or root of how they are calculated? I hope to open up and explore the black boxes of many such issues.

Many people are put off by mathematical technical terms such as the calculus of variations in the problem of quickest descent, Taylor expansion in Machin's formula and Pi, group theory in Burnside's lemma and the theory of cyclotonic equations in Gauss' method of constructing a regular heptadecagon. But through this book we can gradually understand the beauty of maths by trying to discover how to view these problems with interest and find fun in studying them.

I also discuss some of the cultural issues surrounding maths: the huge popularity in 2005 of the number game Sudoku, which is especially liked by European people; why Japanese like odd numbers and Westerners emphasize even numbers, which can possibly be explained according to the principle of Yin-Yang in Chinese philosophy; the difference between Japanese and European architecture which can be explained mathematically in terms of curves and straight lines; and the various methods across the globe for starting to count at "one" with either the index finger, thumb or little finger. All of these differ according to region, ethnicity and historical period. I hope that your curiosity is awakened by the connections between maths and culture as mine was.

This collection of 30 articles addresses just some of the aspects of maths that I enjoy very much. I hope this book will be enjoyed not only by readers who like maths, but also by those who find it difficult and want to gain some insight into the world of maths and logical thinking.

Yutaka Nishiyama

July 2013

Contents

Preface i

CHAPTER 1 Why Do Boomerangs Come Back? 1

CHAPTER 2 Five Petals: The Mysterious Number "5" Hidden in Nature 13

CHAPTER 3 An Elegant Solution for Drawing a Fixed Point 27

CHAPTER 4 Stairway Light Switches 43

CHAPTER 5 The Mathematics of Fans 55

CHAPTER 6 The Mathematics of Egg Shape 65

CHAPTER 7 What's in a Barcode? Duplicated Combinations 75

CHAPTER 8 Building Blocks and Harmonic Series 85

CHAPTER 9 Playing with Möbius Strips 95

CHAPTER 10 A General Solution for Multiple Foldings of Hexaflexagons 105

CHAPTER 11 Turning Things Inside Out 117

CHAPTER 12 Miura Folding: Applying Origami to Space Exploration 127

CHAPTER 13 The *Sepak Takraw* Ball Puzzle 137

CHAPTER 14 Increasing and Decreasing of Areas 149

CHAPTER 15 The Mysterious Number 6174 157

CHAPTER 16 Numerical Palindromes and the 196 Problem   169

CHAPTER 17 From Oldham's Coupling to Air Conditioners   179

CHAPTER 18 Measuring Areas: From Polygons to Land Maps   189

CHAPTER 19 Sicherman Dice: Equivalent Sums with a Pair of Dice   199

CHAPTER 20 Unexpected Probabilities   209

CHAPTER 21 Opening the Black Box of Random Numbers   219

CHAPTER 22 Calculating $\sqrt{2}$ 229

CHAPTER 23 The Brachistochrone Curve: The Problem of Quickest Descent   239

CHAPTER 24 Machin's Formula and Pi   251

CHAPTER 25 Burnside's Lemma   261

CHAPTER 26 Gauss' Method of Constructing a Regular Heptadecagon   271

CHAPTER 27 *Sudoku*: The New Smash Hit Puzzle Game   285

CHAPTER 28 Odd and Even Number Cultures   295

CHAPTER 29 Cultures of Curves and Straight Lines   305

CHAPTER 30 Counting with the Fingers   313

CHAPTER 1
Why Do Boomerangs Come Back?

Abstract: After touching on the three most common misconceptions regarding boomerangs, the author goes on to explain why boomerangs are crescent shaped. The author explains, using the principle of precessional motion, why boomerangs turn leftwards and why they fall sideways. He performs a comprehensive analysis through the "right-hand rule," using the example of a gyro top. He also explains how to make and fly the boomerang he invented — one that can be flown inside a room and returns correctly.

AMS Subject Classification: 00A79, 70E05, 97A20
Key Words: Moment of inertia, Lift force, Left turn and sideways fall, Torque, Couple, Precessional motion, Gyroscopic effect, Right-hand rule

1. Certain Misconceptions

The space available here is far too limited for one to discuss the subject of boomerangs at length. While I would like to offer a detailed explanation on such issues as the ways to make boomerangs and their throwing techniques, methods for control and coordination, and boomeranging as a competitive sport, I have no option but to omit all such practical aspects here.

My initial research in 1978 [4] was by no means exhaustive; however, by referring to the research conducted by Felix Hess [2] and Jearl Walker [3], both the theory and technique have progressed to a large extent, and it can now be said to be very near completion [5]. In this article, I put forward explanations focusing on reasons why boomerangs fly back.

There are three commonly perceived causes as to why a boomerang comes back:
  1. It is crescent shaped
  2. Wind power forces it back
  3. It follows the same principle as the curving of a spinning baseball

First and foremost, it is a misconception to believe that a boomerang flies back because it is shaped in a crescent form. A look at the boomerang models that are presently popular should explain why. There are boomerangs on the market that have three or four wings. There are even some that are shaped like a kangaroo or seagull. They are all constructed to allow them to come back to the thrower. The issue of whether all these can be called boomerangs often comes under debate. For a cultural anthropologist, a crescent-shaped object would constitute a boomerang regardless of whether it comes back to the thrower; a natural scientist, on the other hand, would call any object that comes back to the location of the thrower a boomerang regardless of the shape or number of wings.

Second, that a boomerang comes back due to the presence of wind is also a misconception. For this, it is enough to understand that a boomerang will come back even inside a room where there is no wind. Wind is not a direct cause for its coming back. It is the presence of air and not wind that causes a boomerang to come back. It will return as long as there is air present.

The third and final supposed cause, that the curve principle causes the return, is also a misconception. When a baseball is made to spin with force at the time it is thrown, it either curves or shoots, causing a horizontal change in the path. This is called the Magnus effect after the name of its discoverer. According to the work of Shinji Sakurai [1], a curveball thrown at an initial velocity of 30 m/s will reach the home base 18 m away with a horizontal shift of 40 cm. This change in direction by the Magnus effect is extremely small. The protagonist in the comic book *Kyojin no Hoshi* (*The Star of Giants*) is shown throwing a baseball from inside a room and catching it on its way back, but this is pure fiction.

2. Why Is a Boomerang Crescent Shaped?

The first weapons of early hunting tribes were stones and sticks. What happens when these are thrown into the air? A stone will fly in a parabolic movement, but that won't happen with a stick. Even if thrown with great force, a stick will only twirl in the air before falling to the ground and it won't go far. That is because a stick rotates. A stone can work as a material particle, but a stick has to be considered a rigid body.

Rotary motion is governed by the property that an object at rest remains at rest and an object in rotation always remains in rotation.

This is a well-known law of Newton. Moment of inertia, which is calculated by multiplying mass by distance squared, is a physical quantity that signifies the ease or difficulty of rotation.

There are three coordinates that form the basis for the center of gravity of a board. The direction of the width forms the x-axis, the length the y-axis and the thickness the z-axis. Each of the three axes has its own moment of inertia surrounding it; in terms of comparative size, the moment of inertia around the z-axis is the greatest, followed by the x-axis and finally the y-axis.

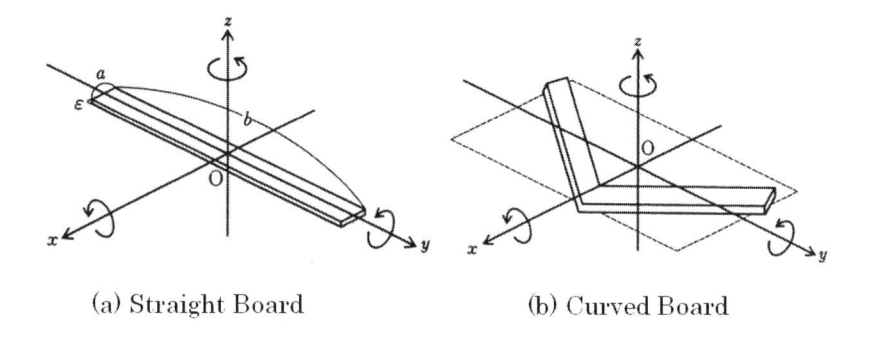

(a) Straight Board   (b) Curved Board

Figure 1: Three Moments of Inertia

The significance of this is that rotation around the z-axis is difficult to bring to rest if in motion and difficult to set into motion if at rest; the y-axis rotation is the easiest to bring to rest or to set into motion.

If the z-axis around which the moment of inertia is largest is made to rotate, with any disturbance in the air stopping the rotating motion from being maintained, it will immediately shift to rotation around the y-axis where the moment of inertia is the least. Thus, the board has to be curved to overcome this drawback. By curving the board, its apparent length is reduced and the width increases. Also, width-wise, air resistance increases, making it difficult for the y-axis to rotate (Figure 1).

Another interesting point concerns shifting the center of gravity of the board. If a board that has been shaped in a crescent form is thrown into the air, a cavity forms in the center, making it much like a doughnut. This cavity balances the air pressure on the top and bottom surfaces of the board, thus increasing the stability of flight (Figure 2). It is for these reasons that boomerangs are curved in a crescent shape.

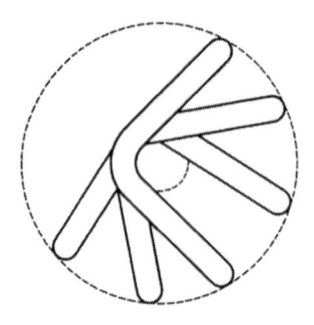

Figure 2: A Cavity is Formed

3. Lift Force Alone Will Not Make a Boomerang Fly Back

A boomerang comes flying back. In other words, flying and coming back are the key parameters. First and foremost, lifting power or lift force is necessary to make a boomerang fly.

Lift force is created by the flow of air; this occurs in two ways. The first depends on the shape in cross-section of the wing. This is called the *aerofoil cross-section*. The shape of the cross-section of an airplane wing shows that whereas the upper surface is convex, the lower surface is flat. When a wing of such a shape moves in the air at a certain speed, the airflow over the top surface is slightly distorted, whereas it passes straight and undisturbed under the lower surface. Because the distance crossed on the distorted upper surface is longer, the speed increases as compared to the air flowing under the lower surface. The surface where air speed is faster experiences lower air pressure as compared to the slower side; that difference in pressure creates a push from the lower surface toward the upper surface. This push is the lift force and is known as Bernoulli's principle.

In that case, would it be correct to say that there is no lift force generated if the cross-section of the wing doesn't show convexity of the upper surface? This is not so. As seen when throwing a celluloid sheet into the air, lift force is generated even in a flat plastic sheet. Second, lift force is generated here by a certain degree of angle of attack against the wind that is present. *Angle of attack* is the angle formed between the base line of the cross-section of the wing and the direction of flight, that is, the direction of flow. It is necessary that this angle be between $5°$ and $10°$.

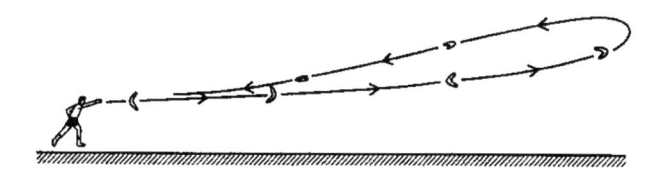

Figure 3: Vertical Throwing

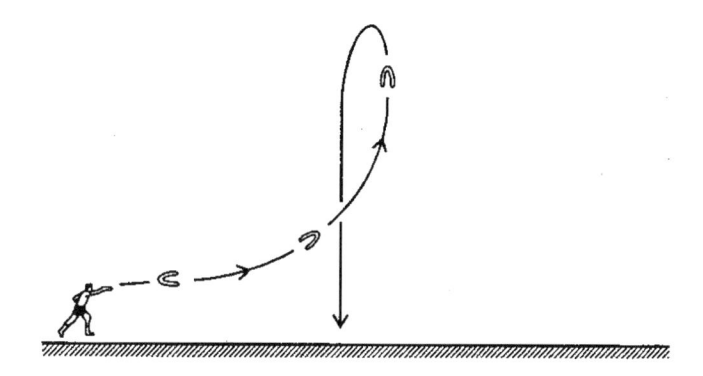

Figure 4: Horizontal Throwing

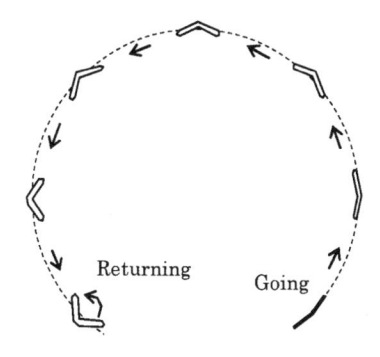

Figure 5: Left Turn and Falling in a Horizontal Position (Seen from Above)

4. Boomerangs Make a Left Turn

What is important when throwing a boomerang is to "throw vertically" and to give a "spin." Ninety-nine percent of failures in boomerang throwing result from "horizontal throwing." If the relation between the way of throwing and the trajectory of the boomerang when a right-handed person throws a boomerang is observed, it is understood that the following three phenomena occur (Figures 3, 4 and 5).

1. A vertical throw results in the boomerang turning left and coming back.

2. A horizontal throw results in the boomerang rising steeply and falling straight down.

3. A vertical throw results in the boomerang making a sideways fall and returning in the end in a horizontal position.

Here, I shall begin by explaining the reasons for the above. Let us first look at the relation between a boomerang's forward motion and spin. Let us suppose that the boomerang shown in Figure 6 is proceeding away from the thrower at a forward velocity of 100 km/h and a rotational velocity of 20 km/h. If the two wings are compared, the wing on top experiences forward velocity of 100 km/h and additional rotational velocity of 20 km/h, making the total speed 120 km/h. The bottom wing, on the other hand, faces the direction directly opposite from that of the spin, and thus the speed is the forward velocity of 100 km/h minus the rotational velocity of 20 km/h, making it 80 km/h. The respective air speeds now are 120 km/h and 80 km/h, accounting for a difference of 40 km/h.

What does the difference in speed affect? It affects the extent of lift force. The lift force is increased in high speed and consequently falls when the speed drops. Due to this difference in lift force, the force to turn the top edge toward the leftward direction, in other words in a counter clockwise direction, comes into play. This rotational force is known as *torque*.

Because torque works in a counter clockwise direction, one would assume that the top edge should land up, being turned leftwards when seen from the perspective of the boomerang thrower. I thought so too at first. But this is not the case. The truth is that a completely unlooked-for phenomenon occurs.

When torque comes into play to make the rotating surface rotate in a counter clockwise direction for the rotation axis of the boomerang, an additional force comes into play to allow the boomerang to maintain its own rotation axis. This is the force of precession. This precession force works on the third axis that lies orthogonal to both the boomerang's rotation axis and the torque axis. Such a motion is known as *precession motion* or the *gyroscopic effect*. The force of precession causes the direction of movement of the boomerang to be turned to the left, and the boomerang returns as a result.

It is now understood that a vertical throw causes the path to turn left. Why, then, does a horizontal throw cause the path to rise steeply? This is also due to precessional motion. If you tilt your head rightwards to a 90-degree angle, you would probably realize that the phenomenon for these is the same. If one presumes that the boomerang is advancing straight along the horizon, the left turn would mean rising upwards. That is why throwing a boomerang horizontally makes it rise steeply before coming to a swift fall.

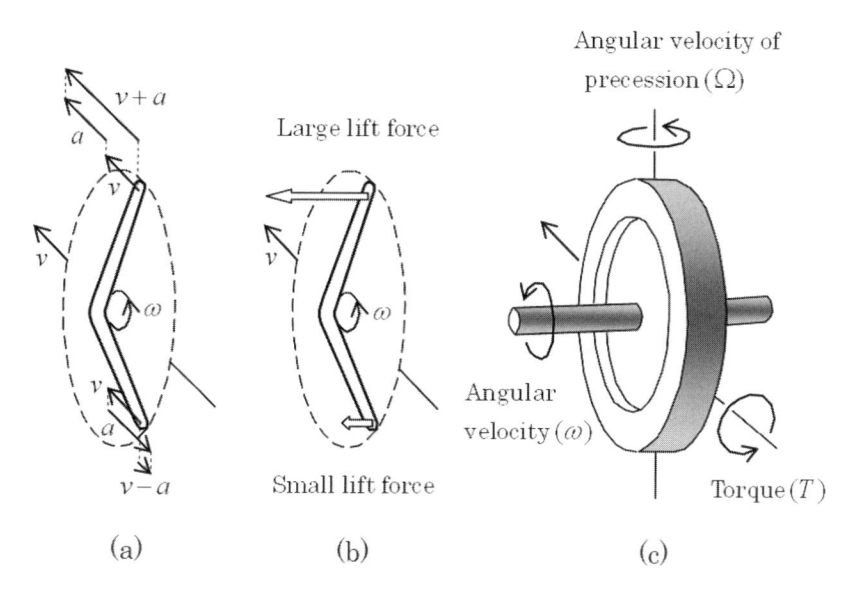

Figure 6: Left Turn Caused by Precession
Adapted by the author from Hess [2].
Note: This could be understood by applying forward velocity of 100 km/h and rotational velocity of 20 km/h.

5. Right-Hand Rule

A boomerang, along with making a left turn when thrown vertically, also makes a sideways fall and returns in a horizontal position. Difference in

8   CHAPTER 1

the lift force between the top and bottom of the wing forms the force to left-turn the boomerang, whereas the difference in the lift force between the front and back of the wings forms the force to make the boomerang fall sideways (Figure 7).

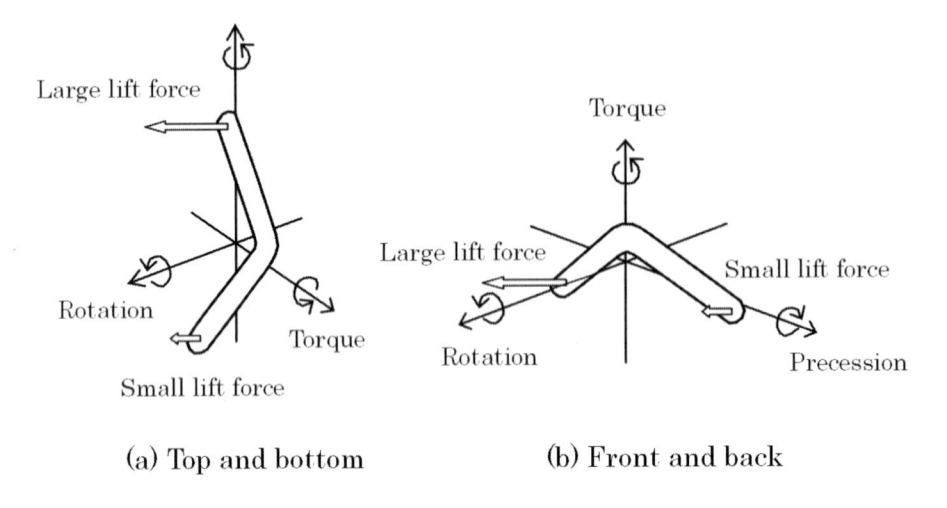

(a) Top and bottom

(b) Front and back

Figure 7: Difference in Lift Force between Top and Bottom and between Front and Back

The phenomenon of left-turning and falling sideways can be understood together if one uses a gyro top.

Spin a gyro top with great force and, as shown in Figure 8, hold the top softly using the thumb and middle finger of the right hand and slide the thumb toward the left and the middle finger toward the right. On doing this, the inner disc does not show signs of falling in a given direction but instead changes direction toward the left side. This is the left-turning of a boomerang. In the same way, hold the top as shown in Figure 9 and slide the middle finger toward the left and the thumb toward the direction of the right side. On doing so, the disc tries to change from a perpendicular position to a horizontal position. This is the boomerang's phenomenon of falling sideways.

To understand the relation between the axes of precessional motion, adjust the thumb, index finger and middle finger of your right hand so that they are orthogonal to each other. As such, if the rotation axis of the top is the middle finger, the torque axis from the difference in lift force is the index finger, and the thumb signifies the precession axis. This is shown in Figure 10, including the direction of rotation. Thus, we understand that in the left-turning caused by the difference in lift force between top and bottom, and in the sideways fall caused by the difference in lift force between front and back, the precession axis is maintained by this right-hand rule [3-5].

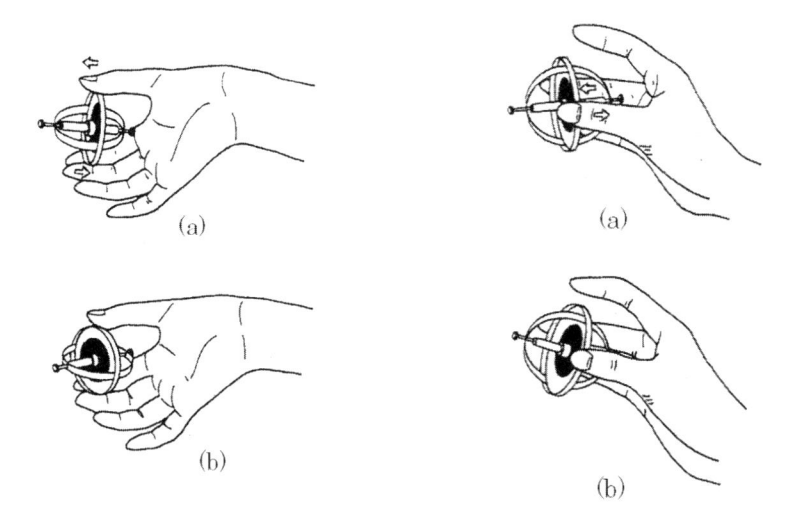

Figure 8: The Plane of Rotation Changes Direction if a Force Couple is Applied from the Top and Bottom

Figure 9: The Plane of Rotation Becomes Horizontal When a Force Couple is Applied from the Front and Back

6. Paper Boomerangs That Come Back Accurately

Here, I shall give a general outline about how to construct and fly paper boomerangs as illustrated in Figure 11. If you follow the procedures written below, almost anyone is guaranteed to be able to catch a boomerang successfully [6].

Create a pattern along the scale illustrated in Figure 11 by making enlarged photocopies. Trace this on an advance copy (using cardboard 0.5-0.7 mm thick) and cut it out neatly using scissors. Put a sign or note saying *front* so as not to mistake the front and back sides. It is particularly important to keep the difference between the front and back sides in mind. Place a ruler along the dotted line and indent the line deeply three or four times using a ballpoint pen. This is done to make it easy to bend or turn along these lines.

10 CHAPTER 1

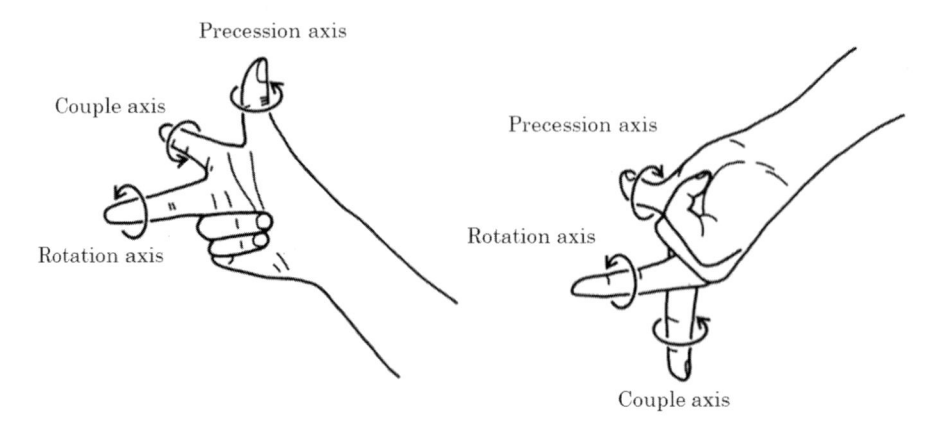

(a) Explanation of left turn   (b) Explanation of sideways fall

Figure 10: Right-hand Rule

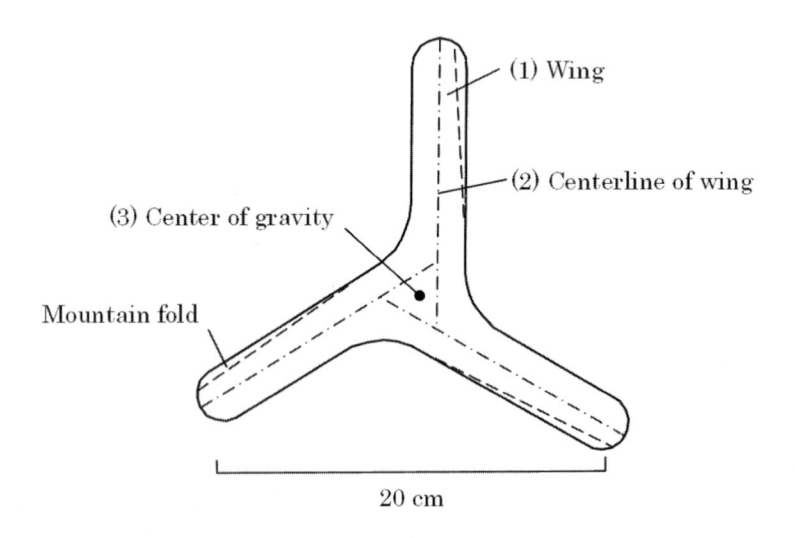

Figure 11: Swept-back Wing Boomerang

With the front side facing upward, make mountain folds along the three wings at an angle of 30 degrees (left-handed people should make valley folds). Place the front side on top and warp all three wings a little upward. Do it so that when placed on an even surface, the tips of the wings protrude upward a little (Figure 12).

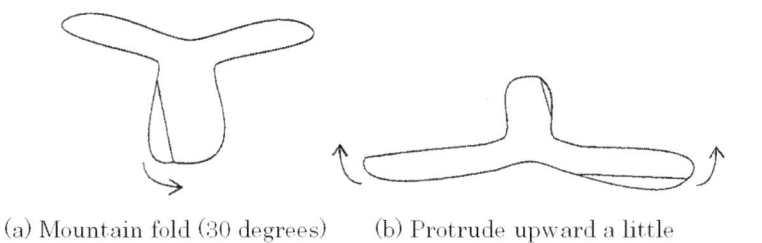

(a) Mountain fold (30 degrees)   (b) Protrude upward a little

Figure 12: How to Coordinate

Place it so that you face the front side, and hold it so that you pick up the tips of the wing with your thumb and index finger (left-handed people should do this facing the back side). Place the wings vertically as though flying a paper plane or throwing darts, and throw it in a straight line at eye level. Make sure you don't throw it pointing upward toward the ceiling. Give a snap to your wrists to give a spin to the boomerang. The swifter the spin, the better the boomerang will come back to you (Figure 13).

The boomerang turns leftward at eye level (counter clockwise when seen from above) and returns (clockwise trajectory in the case of left-handed throwers). The distance flown is 3-4 m, and the time in flight is 1-2 s. Because the boomerang returns in a horizontal position, open your hands out around 30 cm apart, and catch it fast using the palms of your hands. It is dangerous if a boomerang hits the face; make sure that there are no people nearby when throwing the boomerang.

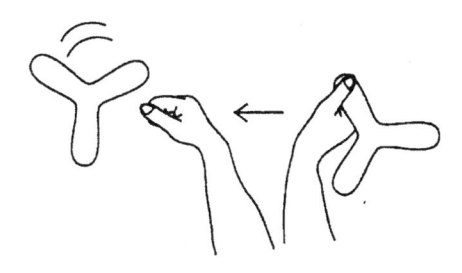

Figure 13: Throwing Technique

References

[1] S. Sakurai, *Nageru Kagaku* [*The Science of Throwing*], Tokyo: Taishukan, (1992).

[2] F. Hess, The aerodynamics of boomerangs, *Scientific American*, 219(5), (1968), 124-136.

[3] J. Walker, The amateur scientist: Boomerangs! How to make them and also how they fly, *Scientific American*, 240(3), (1979), 130-135.

[4] Y. Nishiyama, Boomeran no hikorikigaku [Aerodynamics of boomerangs], *Sugaku Semina*, 17(12), (1978), 34-41.

[5] Y. Nishiyama, The world of boomerangs, *Bulletin of Science, Technology and Society*, 22(1), (2002), 13-22.

[6] Y. Nishiyama, Let's Boomerang! (2007). Instructions in 70 languages can be downloaded.

http://www.kbn3.com/bip/index2.html.

CHAPTER 2
Five Petals: The Mysterious Number "5" Hidden in Nature

Abstract: This article examines why many flowers are five-petaled through the use of a five-petal model that draws insight from the location of cell clusters at a shoot apex, rather than from concepts such as the Fibonacci sequence or the Golden ratio which have been referred to in the past. The conclusions drawn are that flowers are most likely to be five-petaled, followed by six-petaled flowers, and that four petals are unstable and almost no flower can be seven-petaled.

AMS Subject Classification: 00A71, 46N60, 92B05
Key Words: Flower, Cell, Petal, Pentagon, Fullerene

1. Many Five-Petaled Flowers

I am deeply interested in pentagonal forms in the natural world. A hexagon, as seen in bees' nests or snow crystals, is mathematically explained, but no clear explanation is made about a pentagon.

Echinodermata such as sea urchins, starfish and sea cucumbers are five-actinomorphic with a bony plate on the skin and a unique water-vascular system. In other words, they are pentagonal and rotationally symmetrical. The arm of a starfish has strong regeneration power as indicated by the fact that one of its five arms can regenerate immediately. Even more surprising is the fact that one arm can regenerate the remaining four arms [4]. Is something that determines five arms strongly coded for in DNA?

Using a compass or ruler, a regular pentagon can't be drawn as easily as a regular triangle, tetragon or hexagon. Although we can draw a regular pentagon with a protractor, by way of a 360° central angle which is divided into five, starfish don't use a compass or ruler, nor do they have mathematical knowledge. How can such a primitive aquatic creature draw a regular pentagon so easily?

The number "5" seen in sea urchins and starfish is also observed in plants. Flicking through the *Illustrated Guide to Plants (Shokubutu no Zukan)* (published by Shogakukan), many five-petaled flowers are found: spring flowers such as cyclamen, pansy, gypsophila, ume (Japanese apricot), cherry, azalea and peach; summer flowers such as morning glory, bell bind and oleander; as well as autumn flowers such as cotton rose, balloon flower, gillyflower and gentian produce five-petal flowers. Agricultural produce such as watermelon, melon, pear and apple also have five petals.

Having said that, there are also exceptions: calla has one petal; iris has three petals; daphne, dogwood and fragrant olive have four petals; and lily, narcissus and orchid have six petals.

As for the one-petaled and four-petaled flowers among these exceptions, some theories say that what looks like a petal is actually a sepal. First, calla belongs to the arum family, and what appears to be a white petal is in fact a bract with small flowers on a thick axis inside. This bract is sometimes called a spathe, as it looks like the halo or flames often seen on statues of Buddha. Second, with respect to four-petaled flowers, the spring flower daphne has four sepals, not petals. The same thing applies to the autumn flower of fragrant olive with its yellow blossoms.

2. Trimerous, Tetramerous and Pentamerous Flowers

Makino [6] neatly classifies all plants according to kingdom, division, class, order, family, genus and species. His book specifies the number of sepals, petals, stamens and pistils of all the families. A flower diagram shows the pattern of locations and distributions of these to help readers to understand the structure of a flower.

The classification of trimerous, tetramerous and pentamerous flowers is based on a flower's components. If flowers have three (or multiples of three) sepals, petals, stamens and pistils, they are called trimerous flowers. Many monocotyledons such as lily, iris and spiderwort belong to this category. Similarly, flowers with four (or multiples of four) components, such as Japanese laurel and evening primrose, are tetramerous, and those with five (or multiples of five), such as azalea and morning glory, are pentamerous.

Flowers that belong to the same family have the same number of petals. All you have to do in order to know the number of petals of plants is research them at the family level. The result of such research is described below. Let's take the spermatophyte division as an example; it has 219 families. The division is divided into the gymnosperm subdivision (13 families) and the angiosperm subdivision (206 families). The gymnosperm subdivision has no sepals or petals, and is classified as zero-petaled. The angiosperm subdivision branches off into the monocotyledon class (35 families) and the dicotyledon class (171 families). The monocotyledon class includes the iris lily families, and many families in the class are three-petaled or six-petaled (trimerous). They have no sepals, and are counted by tepals. The dicotyledon class is split into the choripetalae subclass (125 families) and the gamopetalae subclass (46 families). Many families in the choripetalae subclass are five-petaled (pentamerous) like the rose, mallow and violet families, or four-petaled (tetramerous) like the mustard and dogwood families. The gamopetalae subclass that includes the heath and morning-glory families has many five-petaled (pentamerous) families.

Table 1 categorizes the 219 families in the spermatophyte division according to the number of petals.

Number of Petals	Number of Families	Percentage
0	38	17.4%
1	2	0.9%
2	6	2.7%
3	13	5.9%
4	38	17.4%
5	84	38.4%
6	24	11.0%
More	7	3.2%
Unknown	7	3.2%
Total	219	100.0%

Table 1. Classification of the Spermatophyte Division by Number of Petals (All)

Table 2 shows just the three- to six-petaled families. Those families with no petals are included as long as the number of sepals or bracts can be counted. As a result, it is found that the largest number of families is five-petaled and that five-petaled flowers belong to the dicotyledon class, the angiosperm subdivision, which is an advanced plant community from the perspective of evolution theory.

Number of Petals	Number of Families	Percentage
3	13	8.2%
4	38	23.9%
5	84	52.8%
6	24	15.1%
Total	159	100.0%

Table 2. Classification of the Spermatophyte Division by Number of Petals (3-6 Petals)

3. Chrysanthemum: A Five-petaled Flower

We are familiar with chrysanthemums. Again, according to the *Illustrated Guide to Plants*, the aster family accounts for the largest share of 9%, with its 135 species among the total of 1495 species. Members of the aster family include dandelion in spring, sunflower in summer and cosmos in autumn. I had mechanically classified the aster family as multipetaled (seven or more), but it is actually five-petaled.

A flower in the aster family consists of a ray flower and a tubular flower, the latter surrounded by the former. The ray flower is a coalescent five-petaled flower. That is, the four of the five original petals had been degenerated with one ray petal remaining. The tubular flower at the center is an aggregate flower with hundreds of small, densely packed flowers.

Cosmos looks like an eight-petaled flower but these eight petals are actually eight flowers. You can see this by looking at its tubular flowers through a magnifying glass with a power between 10 and 15. Each one of the densely located small flowers splits into five at the edge. Its appearance is similar to the flower of a balloon flower and obviously five-petaled.

The flower of a tare in the legume family is tubular at the bottom and divided into five at the edge. Its one flag petal, two wing petals, and two keel petals constitute the corolla (the entire group of petals). The flowers of the legume, mint and violet families are symmetrical if horizontally observed. Such a flower is called a zygomorphic flower. In any case, they are five-petaled.

Okra, which is edible, has the shape of a regular pentagon. It is also called *America neri* in Japan and belongs to the mallow family, in the *Hibiscus manihot* genus (*Abelmoscus*). The mallow family belongs to the mallow order (Malvales), the choripetalae subclass, the dicotyledon class, the angiosperm subdivision, and it is pentamerous. Its sepals, petals and stamens can be seen from outside, but its ovary can't because of the pentagonal shape of the fruit. The ovary, located below the pistil, is divided into five cells, which makes the fruit pentagonal. The cross-section of the fruit of a pear is pentagon. Like okra pear belongs to the rose family, which is pentamerous.

4. What is a Flower?

Let us get an idea about what a flower is in terms of its shape, referring to Hrara [2]. A plant's organs are the root, stem and leaf. Roots constitute the part of the plant body that is underground: they exist to support the plant and to absorb water and inorganic salts. Stems make up a part of the plant body that is aboveground, and are designed to support the plant and disseminate matter. Leaves line up in a regular manner around the stems and carry out photosynthesis. Through its leaves, the plant actively provides oxygen in exchange for carbon dioxide from outside, and transpires.

A unit consisting of a stem and leaves regularly arranged around the stem is called a shoot in botany. One example of a shoot is a branch. Inflorescences, buds and growths that develop and extend from the stem are also considered shoots.

Many plant flowers consist of sepals, petals, stamens and a pistil, which is formed from one to several carpels, equivalent to leaves, sticking to one another. Sepal, petals and stamens are also regarded as modified leaves. Therefore, a flower can be considered to be a metamorphosed short stem with variations of leaves regularly surrounding it (Figure 1).

A flower, from the perspective of botany, is an organ that produces a fruit and seed. In other words, a flower is ultimately a reproductive organ.

Some mechanism to determine a five-petaled characteristic may exist in pollen. From this standpoint, I looked at the picture of pollen taken with a scanning electron microscope in the *Compendium of Modern Botany, Vol. 7 (Gendai Seibutsu-gaku Taikei, Di 7 Kan)*. Although various forms of pollen are observed, some of which are spherical-like fertilized eggs and the morulae of sea urchins or starfish, no sign of anything that might decide the number "5" is apparent. The next candidate is the seed, in which the origins of only a cotyledon, plumule and radicle, (i.e., the leaf, bud and root) are found, with no sign of a flower. This is not altogether surprising as the flower doesn't form until later, in the second phase of the plant's growth.

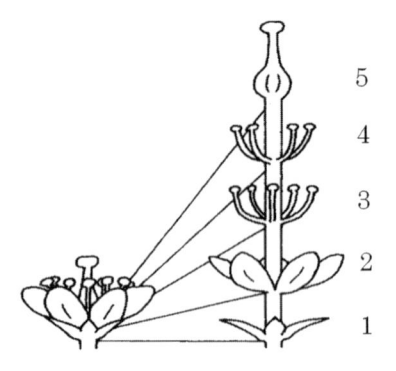

Figure 1: The Flower as a Shoot
(1: Sepal; 2: Petal; 3 and 4: Stamens; 5: Pistil).
Adapted by the Author from Hara [2].

5. Spiral Phyllotaxis and the Fibonacci Sequence

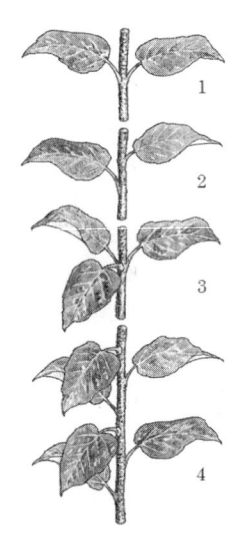

Figure 2: Leaf Arrangement
(1. Opposite, 2. Alternate, 3. Verticillate, 4. Spiral)

The rafflesia found in the tropical rain forests of Sumatra, the largest flower in the world, is five-petaled. I saw a TV documentary on how its flower blooms. I had thought that one petal was followed by a petal next to it when the flower bloomed, but in fact the next petal was skipped and alternate petals open in the first round, with the five-petaled flower fully in bloom in the second round. This order in blooming is confirmed by peeling and observing the petals of a blossom bud pulled off a plant of the tea family.

The order of petal opening is similar to that of leaf arrangement, which is explained later (spiral phyllotaxis). This suggests that a flower consists of metamorphosed leaves with a very shortened stem. By the way, the leaf primordium theory about a flower was originally advocated by Goethe in 1970. This theory is correct on the whole, although it also contains an error in the sense that a leaf bud and a flower bud are different and that all the leaves do not become flowers.

The role of leaves is to photosynthesize. Photosynthesis must make the most of the sunlight available, and leaves are arranged so that, as far as possible, leaves do not cast shade on each other. There are four such arrangements (Figure 2). The first is called opposite, with two leaves at the same height on the stem. The second is alternate, with leaves alternately arranged at different levels on the stem. The third is verticillate, with three leaves at the same height, and the last is spiral, with three leaves at a different heights. When there are many leaves, a spiral arrangement can reduce the casting of shade most effectively. The arrangement of leaves is called phyllotaxis and a spiral leaf arrangement is called spiral phyllotaxis.

An alternate leaf arrangement and a spiral phyllotaxis are further classified by the divergence, the angle formed by the two lines on the sectional view of a leaf arrangement, one line connecting a stem and the center of a leaf and the other line connecting a stem and the center of another leaf immediately above or below it. When the divergence is, for example, one of 144°, the leaf arrangement is called a two-fifths leaf arrangement because 144° is two-fifths of 360°. Many types of leaf arrangement exist, such as a half, one-third, two-fifths and three-eighths. It can be considered that the numerator and denominator reflect the numbers of spirals and leaves in the spirals, respectively. For example, a three-eighths leaf arrangement implies that there are eight leaves in three spirals.

These leaf arrangements of plants can be explained by way of the Fibonacci sequence. The Fibonacci sequence, which is taught in high school mathematics, derives from botanical studies, and is a sequence formulated by the sums of two previous terms, as $1, 1, 2, 3, 5, 8, 13, \cdots$ If a pair of every other figures is picked up with the larger one as the denominator and the smaller one as the numerator, they stand for a certain type of leaf arrangement. For example, 2 and 5 form the two-fifths of a two-fifths leaf arrangement, and 3 and 8 form the three-eighths of a three-eighths leaf arrangement.

Can the Fibonacci sequence explain all leaf arrangements? In response to this question, Hara [1] makes an interesting suggestion: "Such a mathematical relationship is found, it is natural that academic interests deepen in that direction. Arguments on leaf arrangements thus proceeded on the basis of mathematical figures rather than actual botanical observations of the leaf production process on the stem apex, and proceeded in connection with a theory that the Fibonacci sequence had a direct bearing on the golden section. The arguments developed as far as a theory that a leaf primordium on a stem apex came out ultimately at the locations related to the golden section. However, it is not appropriate to discuss this issue only with reference to mathematics, without actual and elaborate observations on how a leaf grows from a stem apex and how an embryotic leaf primordium changes its location secondarily."

I do not deny the Fibonacci sequence theory in every aspect, but it does not sufficiently explain why a flower is five-petaled. Actual observations are required.

6. Cells in the Stem Apex

A growing point is a stem apex or a root apex where cell division occurs, and this is certainly a key to solving the mystery. Specifically, the shape, size and arrangement of cells at a growing point should be looked at.

Hara [2] explains the basic structure of a stem apex, which is also called a shoot apex. A shoot is a unit consisting of a stem and leaves. A shoot apex directly generates a stem and leaves as well as the axil meristematic tissue of a leaf.

The apex meristem of an angiosperm shoot has the structure of a tunica, corpus and cellular structure band (Figure 3). The shoot apex has one to several parallel layers of cells on its surface. This layer structure is the tunica, and a nonlayer structure inside is the corpus. The tunica constitutes the cell layers that repeat vertical divisions, and the corpus comprises cells that make division planes in various directions.

In addition to the tunica and corpus structure, the shoot apex has a cellular structure band, which is divided into three zones according to the nature of the component cells. The uttermost end of a shoot apex is called the central band. The zone surrounding the central band is the peripheral meristem, and the zone below the central band, which is surrounded by the peripheral meristem, is the pith meristem.

In the cellular structure band, the central band draws attention. Cells in the central band are large and near circular, and they stain

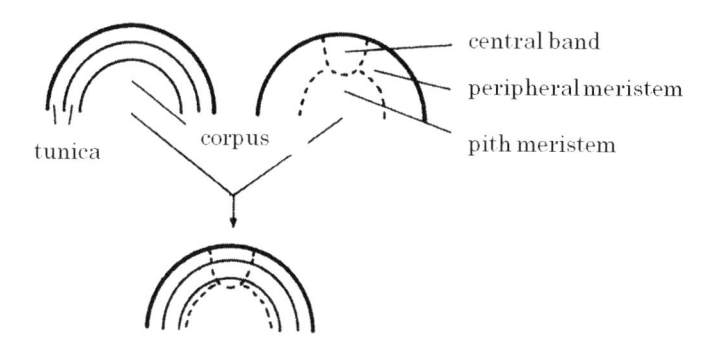

Figure 3: Tunica, Corpus and Cellular Structure Band at the Shoot Apex
Adapted by the author from Hara [1].

weakly and have many vacuoles. In particular, its less frequent cell division is surprising as it is located at the center of the meristem. Cells in the peripheral meristem surrounding the central band are active, and leaf primordia are generated at the part of the peripheral meristem close to the central band.

Major changes take place at the shoot apex when it enters the reproductive phase from the vegetative phase. The cell arrangement of the tunica and corpus as well as the structure of the cellular structure band are gradually lost during this conversion phase, and several layers of cell are generated on the surface of the shoot apex. Cells in this part have a dense cell layer, and are the sites of active cell division.

In the formation of flowers, a reproductive shoot apex generates the primordia of a floral leaf, usually in the order of the primordium of sepals, petals and carpels. When carpel primordia are made up, the reproductive shoot apex completes its division and becomes a part of the carpels to end it [5].

Figure 4 illustrates the growing points as explained below. A stem edge, as seen in its cross-sectional view (Figure 4(a)), can be divided into dome-like cells and lateral parts where small leaves are being formed. The more three-dimensional Figure 4(b) shows the state of vegetative reproduction, in which leaves are generated one after another. On the other hand, Figure 4(c) shows an apical part during the early flower bud formation period. The formation of the flower bud occurs at a centric part, where the growing part of the leaf premordium changes into the flower premordium. Though only by a very small degree, the site of leaf formation and the site of flower formation are disjointed at the apical part.

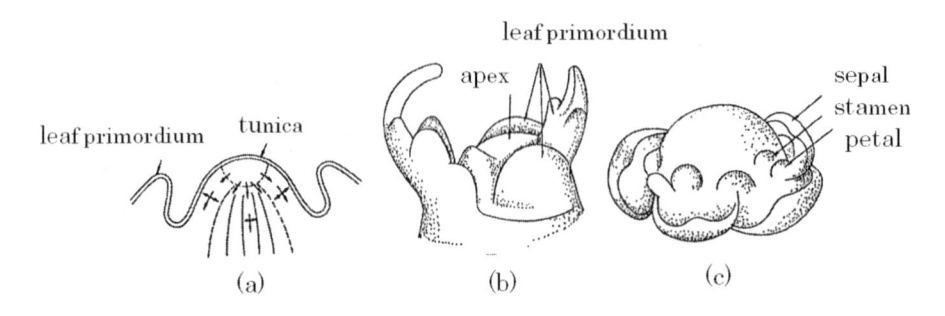

Figure 4: Growing Points of a Flower
(a) Cross-sectional View; (b) Three-dimensional View; and (c) Flower Primordium
Adapted by the author from [5].

7. Fullerrene C_{60}

The 1996 Nobel Prize in Chemistry was awarded for the discovery of a soccer-ball-shape molecule, C_{60}. The prize winners were Robert F. Curl Jr., Richard E. Smalley and Sir Harold W. Kroto. Figure 5 shows the molecule model of fullerene C_{60} and a soccer ball. Carbon atoms are placed at vertexes of the polyhedron. There are 12 pentagonal aspects and 20 hexagonal aspects, with one pentagon surrounded by 5 hexagons. This carbon-only structure, which is perfectly symmetrical, proved to be a totally new concept in molecule structures. The name fullerene is an abbreviated form of Buckminster Fullerene, after the famous architect Richard Buckminster Fuller who invented the geodesic dome structure.

What is mathematically interesting is the fact that a graphite sheet with 60 atoms, according to the Euler's formula, cannot form a closed ball. Similarly, a familiar wire fabric with hexagonal elements cannot form a ball. This is easily proved by Euler's polyhedron theorem. The theorem maintains that $V + F = E + 2$ (V : number of vertices, F : number of faces, E : number of edges). As studies on fullerene C_{60} have continued, soccer-ball-like or multiple-layered molecules with more than 60 carbon atoms as well as "carbon nanotubes," tube-shaped clusters of carbons, have been found one after another. Carbon nanotubes, also called buckytubes, were discovered by Sumio Iijima, a researcher at NEC Corporation, in 1991. If fullerene with the same diameter as C_{60} is extended, the length is about $61 \mathring{A}$ in the case of C_{500}. The noteworthy discovery about nanotubes is that some of them include seven-member rings. While five-member rings close fullerene in a ball shape, seven-member rings extend fullerene. Seven-member rings occur on a saddle-like curved surface (negative curved surface ratio).

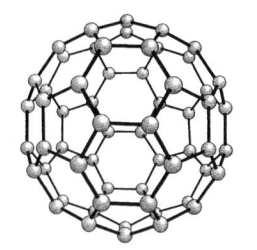

Figure 5: C_{60} Molecule Model and a Soccer Ball

8. Cell Arrangement and Five Petals

The stem apex has a dome-like shape. How are cell clusters arranged in this dome-like shape? The molecule called a carbon nanotube, which is a variation of fullerene, has a long, slender tubular shape reminiscent of a plant's stem. The edge of the nanotube is convex, and has five-member rings. I infer that a stem is similar to a nanotube.

A petal is formed from thousands of cells, not from one cell. The cell cluster that forms the petal is located at the stem apex, and its arrangement certainly has to do with the form of the flower. If the arrangement is important, questions such as how the arrangement develops and what the optimal arrangement is arise. From such a perspective, I have the feeling that it's not difficult for plants to choose five petals; in fact, the choice of four petals is unstable and difficult.

Let us suppose that a cell cluster at the stem apex is a ball (or a hexagon) [3]. In the first place, it is common knowledge that the optimal shape to cover level ground is a regular hexagon. Hexagonal shape filling by circular discs has fewer openings than square shape filling. Arranging 10-yen coins as if the coin were a hexagon, as shown in Figure 6(a), makes the fewest openings, while arranging them as if the coin were a square, as shown in Figure 6(b), makes more openings, which allows the coins to move. In this sense, hexagonal shape filling is the optimal arrangement. It is well known that the regular hexagons seen in bee hives and snow crystals are connected with this. On the other hand, flower petals choose five, not six, in many cases.

(a) Hexagonal shape filling

(b) Square shape filling

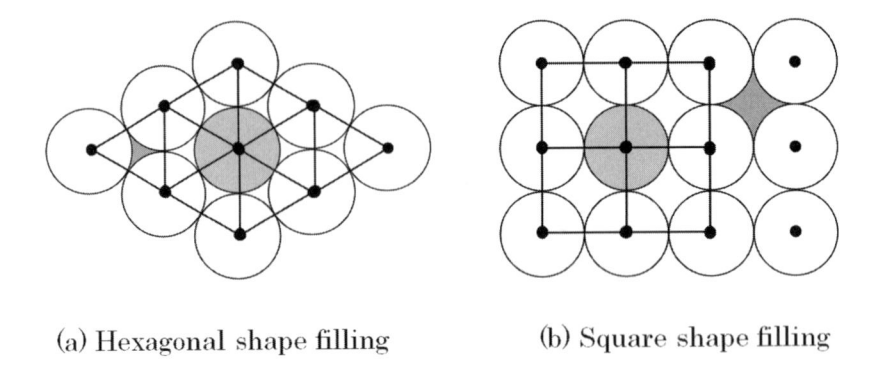

Figure 6: High Density Filling with Circles

A guiding principle in form or pattern generation in nature is an action or binding force of a space or spot. The leaf arrangement of plants can be explained by the Fibonacci sequence or the golden ratio, but plants do not know mathematics. Plants simply try to occupy the largest space they can, and the result of such a tendency happens to be described by those mathematical rules. All the beauty and mathematics is a natural byproduct of a simple growth system that is interactive with a surrounding space [7].

Thus, I present the following cell arrangement models. Figure 7(a) shows six regular hexagons arranged without opening to surround one regular hexagon at the center. As you can see from Figure 7(a), the arrangement is flat. If the number of hexagons is reduced from six to five, the five hexagons forms a shape like an upside-down bowl, convex toward the top and concave toward the bottom, as seen in Figure 7(b). On the other hand, if one hexagon is added, a saddle shape, uneven like waves, is formed by the seven hexagons.

These transformations to a bowl or a saddle depend on the expanding velocity of a surrounding part in relation to a central part. These transformations are attributable to the nature of space. The relatively high growth velocity at the central part creates a dome-like shape, and that at the surrounding part a saddle-like shape. The shoot apex (stem apex) is a growing point, which means that cells at the center are most active, and its shape is convex toward the edge, like a dome.

Cells are, therefore, arranged as shown in Figure 7(b). Five hexagons, or five cell clusters, surround one pentagon, or a stem. These five cell clusters transform into sepals, petals, stamens and ovaries. With this,

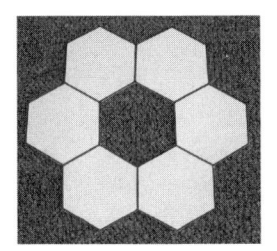

(a) Six Regular Hexagons (b) Five Regular Hexagons

Figure 7: Models of Cell Cluster Arrangements

the mystery of a five-petaled flower is solved. From the same reasoning, we can infer the extremely low possibility of a seven-petaled flower, since a saddle-like shape formed by an arrangement where seven cell clusters surround a stem is unstable and it is not natural for a surrounding part to have higher growth velocity.

A six-petaled flower can be generated from a flat stem apex. If a stem apex is flat, six cell clusters are arranged around a stem as shown in Figure 7(a), and they become sepals and petals, meaning a six-petaled flower is formed. However, it is not natural for the stem apex to be flat, and the possibility of six petals is lower than that of five petals.

A four-petaled flower can be formed when the arrangement of cell clusters at the stem apex is like the square shape filling shown in Figure 6(b). This square shape filling has lower density than hexagonal shape filling, but, given the lower variability of plant cells compared with animal cells, such an arrangement is not impossible to take place. The stem apex is formed as square shape filling and if cell clusters of the stem apex are also arranged in this manner, if we assume that one of them is a stem, there are four neighboring cell clusters around that stem. Accordingly, because these four cell clusters become sepals and petals, the four cell clusters generate a four-petaled flower. The possibility of four petals, however, is lower than that of five petals, since square shape filling is less dense than hexagonal shape filling.

The Greek myths talk about the statue of Daphne. Daphne's arms change to branches and leaves, and her whole body transforms into a tree. If this myth has scientific grounds, it could explain the reason that we have five fingers, by relating the human arm, hand and finger to the plant stem, branch, leaf and flower. Fingers on both hands total 10. Ten fingers might contribute to the decimal system, and be reflected in culture, history, society and ideology. I cannot help feeling that "5" occurs right across the universe.

References

[1] N. Hara, *Shokubutsu no Keitai, Zotei-ban* [*Pattern of Plants, Supplemented and Revised Edition*], Tokyo: Shokabo, (1984).

[2] N. Hara, *Shokubutsu Keitai-gaku* [*Plant Morphology*], Tokyo: Asakura, (1994).

[3] S. Hildebrandt, *Katachi no Hosoku* [*Mathematics and Optimal Form*], Tokyo: Tokyo Kagaku Dozin, (1994).

[4] M. Ichikawa, *Kiso Hassei-gaku Gairon* [*Introduction to Basic Embryology*], Tokyo: Shokabo, (1982).

[5] S. Kaku, *Shokubutsu no Seicho to Hatsuiku* [*The Growth and Vegetation of Plants*], Tokyo: Kyoritsu, (1982).

[6] T. Makino, *Kaitei Zoho, Makino, Shin Nihon Shokubutsu Zukan* [*Makino's New Illustrated Flora of Japan, Revised and Supplemented*], Tokyo: Hokuryukan, (1989).

[7] P. Stevens, *Shizen no Pattern* [*Patterns in Nature*], Tokyo: Hakuyosya, (1987).

CHAPTER 3
An Elegant Solution for Drawing a Fixed Point

Abstract: This paper presents a newly developed theorem for the construction of fixed points under congruent and similar transformations, which differs from existing theorems based on Euclidean geometry. It also explains the random-dot pattern that facilitated the invention of this new theorem.

AMS Subject Classification: 00A09, 51A02, 97G02
Key Words: Fixed point, Construction, Congruent transformation, Similar transformation, Random-dot pattern

1. Elegant Solution

It is probably a common aspiration among mathematicians to find at least one theorem during their lifetime. The theorem presented here is for the construction of fixed points, and the only theorem I can be proud of among those I have invented. I like this theorem, which I personally think is elegant. Suppose that two squares are randomly placed as shown in Figure 1. What is the simplest method to find a fixed point to match the one with the other?

The process of congruent transformation is divided into rotational translation (Figure 2(a)) and parallel translation (Figure 2(b)) or also symmetry translation although we are not aware of such a division in everyday life. These translations may be made in any order.

The combination of parallel and rotational translations can be replaced by one rotational translation, the center of which is called a fixed point in congruent transformation. A general method to find a fixed point is based on Euclidean geometry, as shown in Figure 3. In order to match Square ABCD and Square A'B'C'D', Side AD must be put on Side A'D'. This means that Vertex A and Vertex D are moved to Vertex

Figure 1: Find a Fixed Point

(a) Rotational translation

(b) Parallel translation

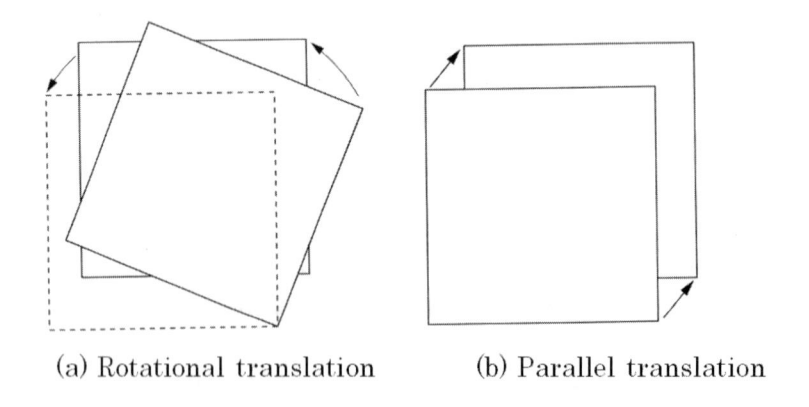

Figure 2: Transformation of Figure

A' and Vertex D', respectively. Point O where the perpendicular bisector of Segment AA' intersects with that of Segment DD' is the center of rotation, or the fixed point, as indicated by the fact that Triangle AOD and Triangle A'OD' are congruent.

Is there a more efficient and elegant method to construct a fixed point? In Figure 4, I take two squares, and draw two straight lines on the square above. If the square above is held by a compass at the point at which the two lines intersect and rotated, the one square completely matches the other. When I tried this experimental movement for the first time, I suspected that the two squares were matched by pure chance. However, repeated attempts from different positions on the paper convinced me of the fact that a fixed point could be constructed in that way. By just drawing two straight lines, a fixed point can be constructed from their intersection [2].

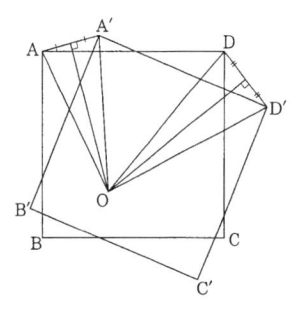

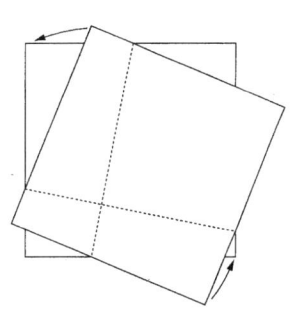

Figure 3: Construction by Perpendicular Bisectors

Figure 4: Elegant Solution

While the existing construction method shown in Figure 3 requires a compass and a ruler to find a fixed point, one feature of my method shown in Figure 4 is that it does not need a compass. The necessary condition of this new way to construct a fixed point is that opposite sides of a quadrilateral are parallel. My theorem therefore applies not only to squares but also to rectangles and parallelograms. If you do not have square paper at hand, you can conduct the experiment with rectangular writing paper or copy paper.

2. Outline of Proof

Let me mathematically prove that the intersection of the two auxiliary lines drawn on Figure 4 is a fixed point. For the sake of the proof, the two pieces of square paper are marked as shown in Figure 5. Points A, B, C and D correspond to each vertex of the square below, and Points A', B', C' and D' to that of the square above. Points P, Q, R and S are the intersections of the sides of the two squares, and Point O is the intersection of Line PR and Line QS.

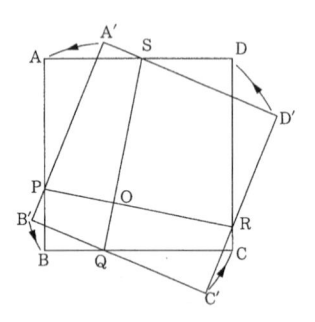

Figure 5: Construction of a Fixed Point

Since it is difficult to give a direct and immediate proof that Point O is a fixed point, I will provide a step-by-step explanation here. As the first step, let me prove "In order to match two lines, a center of rotation must be on the bisector of the intersecting angle." Figure 6 shows Lines l_1 and l_2 intersecting at Point P. Point Q is an arbitrary point on the bisector of the intersecting angle, and Points R and S are foots of perpendicular lines dropped from Point Q to Lines l_1 and l_2. Right-angle Triangles QPR and QPS are congruent since Line QP is common and Angle QPR is equal to Angle QPS. The length of Line QR is therefore equal to that of Line QS. This means that Lines l_1 and l_2 are matched with the rotation centering around Point Q.

The next step is "to determine the conditions required for matching parallel lines." As shown in Figure 7, two parallel Lines l_1 and l_2 with a certain interval intersect with two parallel Lines l_3 and l_4 with the same interval. The intersections are marked as Points E, F, G and H. In order to match Line l_1 with Line l_3, the center of rotation must be on the bisector of an intersecting angle, or Angle E. Similarly, with respect to Lines l_2 and l_4, the center of rotation must be on the bisector of Angle G. Parallelogram EFGH formed by four Lines l_1, l_2, l_3 and l_4 is a lozenge with four sides of the same length, and the bisector of intersecting Angle E exactly corresponds to that of intersecting Angle G. Point Q, an arbitrary point on Line EG, is thus the center of rotation to bring Line l_1 on Line l_3 and Line l_2 on Line l_4 at the same time.

Figure 6: Matching Lines

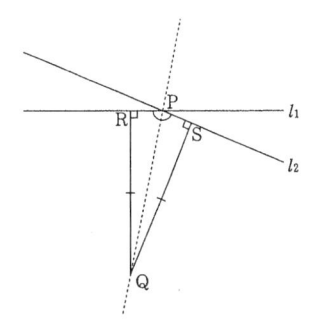

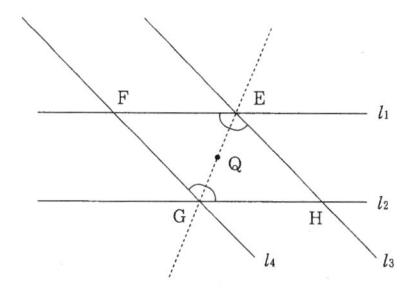

Figure 7: Matching Two Parallel Lines

With the above steps in mind, let us look at Figure 5. Sides AB, DC, A'B' and D'C' in Figure 5 can be regarded as Lines l_1, l_2, l_3 and l_4 in the above steps, respectively. When Sides AB, DC, A'B' and D'C' are extended, they form a lozenge, and Line PR, a diagonal line of the lozenge, bisects both Angle A'PB and Angle DRC'. Any center of rotation on Line PR can therefore match Side A'B' with Side AB and Side D'C' with Side DC. Similarly, any center of rotation on Line QS can match Side B'C' with Side BC and Side A'D' with Side AD. In order to match Square A'B'C'D' with Square ABCD, these two rotational translations must occur simultaneously. Therefore, the Intersection O of Line PR and Line QS is the only center of rotation to match the two squares.

3. Random-dot Pattern

I would like to describe how the simple method to construct a fixed point of congruent squares shown in Figure 4 was invented. As for the existence of fixed points, the fixed point theorem by E. J. Brouwer is widely known. It says that an arbitrary continuous mapping f of space X into itself has at least one fixed point. Fixed points here are defined as point $x \in X$ that satisfies $f(x) = x$. This theorem demonstrates the existence of fixed points, but it is another issue to construct them.

In 1980, I encountered an interesting article by Jearl Walker in *Scientific American* [6]. The article dealt with various experiments using randomly plotted points, and greatly contributed to the invention of my theorem. Prompted by the article, I made a random-dot pattern [3].

The pattern had 2000 points randomly plotted within a 20cm-side square. Computer generated random numbers were used for the (x, y) coordinates of each point, and points were then drawn by a plotter based on the coordinates. Figure 8 shows a random dot pattern plotted in that way.

Further work is required to make this pattern interesting. I printed it on an OHP film, a transparent plastic paper. The film was placed over the original pattern to complete the preparation. When I slightly rotated the film above, a concentric circle surfaced from random dots (Figure 9). This illusion would probably surprise and impress anyone who saw it. You would see only one concentric circle, and no other such circle would surface by any chance. When I counter-rotated the film around the center of the concentric circle, the two patterns, the one on the film and the other on the original paper, coincided exactly. The center of the concentric circle is, in other words, a central axis of the rotation or a fixed point.

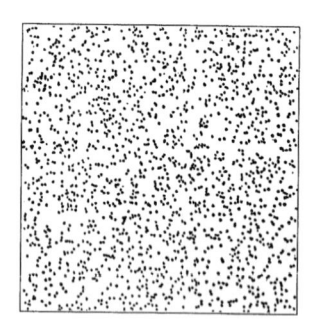

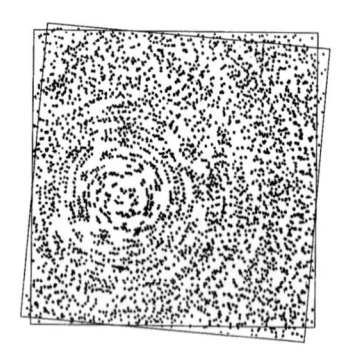

Figure 8: Random Dot Pattern

Figure 9: Emerging Fixed Point

I was fascinated with this enigmatic circle created by random dots, and looked at it all day long. Then, I noticed that the center of the concentric circle corresponded to the intersection of two straight lines connecting the intersection of each side of the two squares. The random assignment of dots is necessary to prevent two or more circles from surfacing, or to make a fixed point visible. To explain the reason, let me use a regular dot pattern with regular placement of dots (Figure 10). The number of dots plotted is 2000 as in the case of the random dot pattern.

The regular dot pattern is printed on paper and OHP film, and the two materials are overlaid with the film slightly rotated. The regular dot pattern produces multiple concentric circles (Figure 11(a)). A wider

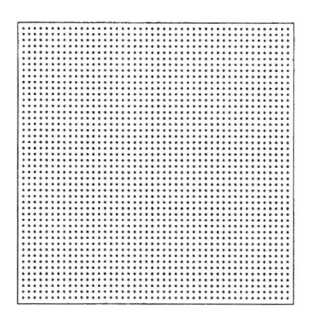

Figure 10: Regular Dot Pattern

rotation angle of the film brings smaller and more concentric circles (Figure 11(b)). Among many centers of these concentric circles, only one center shows a fixed point, and the others are dummies. Regularity in dot placement causes this phenomenon. In order to have only one concentric circle, which demonstrates a fixed point, dots must be plotted randomly.

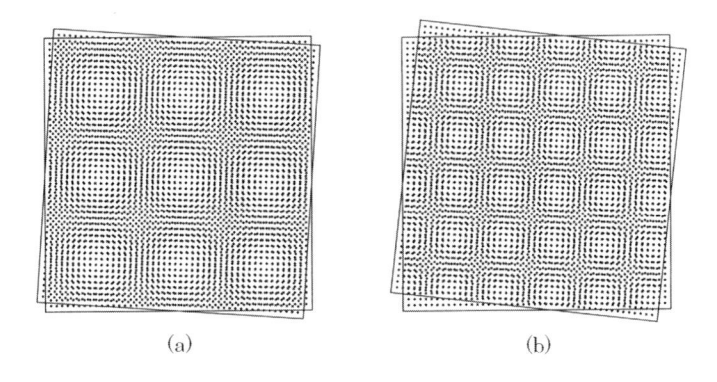

Figure 11: Emerging Multiple Concentric Circles

4. Fixed Points of Circles and Triangles

Since four sides of a square have the same length, there are four types of superimposition alignments of two squares, one of which is shown in Figure 5. Each alignment has a different fixed point. Figures 12(a)-(d) show these four alignments and fixed points. Sides are extended in Figures 12(b) and 12(c) to construct intersections of sides. This expands

Nishiyama's theorem so that it can be applied to when corresponding sides of two squares do not cross each other.

In this way, it becomes clear that squares have four fixed points, which are here identified as P_1, P_2, P_3 and P_4 for convenience. What is their positional relationship? Figure 13 demonstrates that four fixed points can be put on the same straight line. Although such a relationship seems to be mysterious, circumscribing circles in Figure 13 helps with the mathematical understanding of it. Rotational translation of the squares corresponds to rotational translation of their circumscribing circles.

Squares have four fixed points, and regular octagons have eight fixed points. Two regular n-polygons, each of which has n-sides with the same length, can be placed in n-ways, and the number of fixed points is thus n. The fact that these fixed points form a straight line can be clearly understood by drawing circles to circumscribe regular n-polygons, as in the case of squares. A regular n-polygon with infinite n-sides is a circle. Two circles have an infinite number of fixed points, which align on a straight line. In other words, fixed points of congruent circles are on the intersecting line of the two circles, and any point on the line can be a fixed point (Figure 14). Please examine this yourself [4].

A fixed point of congruent triangles can be constructed as follows. I would like to construct Point P, a fixed point of Triangle ABC and Triangle A'B'C' (Figure 15(a)). In order to do this, the two triangles are inverted to form Parallelograms ABCD and A'B'C'D', and then a fixed point for them is constructed (Figure 15(b)). A fixed point for parallelograms can be constructed in the same way as that for congruent squares, because a parallelogram also satisfies the aforementioned necessary condition: parallel opposite sides. Point P, a fixed point of parallelograms constructed in this way, is also a fixed point of triangles. In the case of triangles, however, my construction method is more complex and thus less desirable than the method using perpendicular bisectors that is shown in Figure 3.

5. Similar Transformation

H.S.M. Coxeter provides a detailed explanation of similar transformation [1]. Coxter takes a photo enlarger and a pantograph as examples of enlarging transformation, and maps with different scales as examples of spiral transformation. Figure 16 illustrates maps piled up in the order of scales. A map with a 1/20,000 scale is placed on a larger map with a 1/10,000 scale. A 1/40,000 scale map is subsequently laid over the

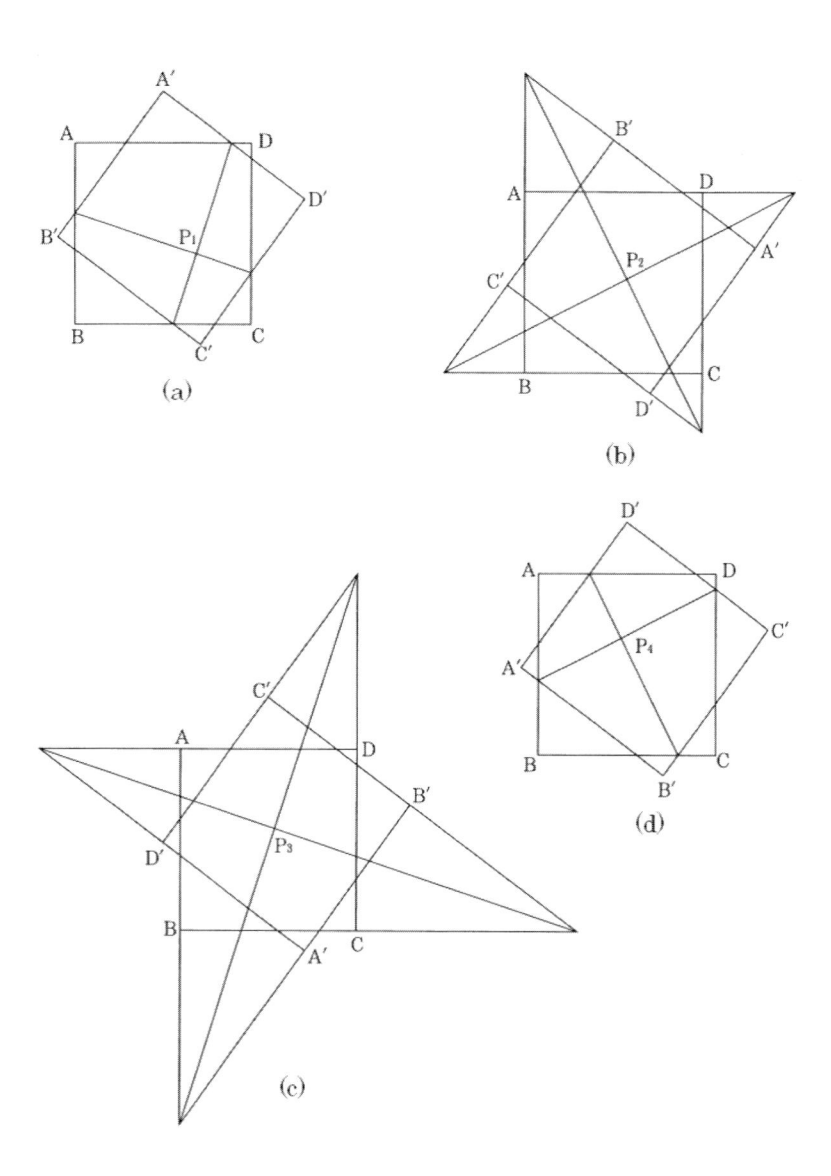

Figure 12: Four Fixed Points

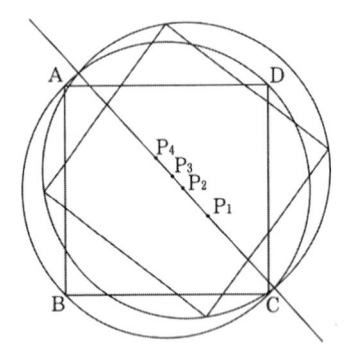

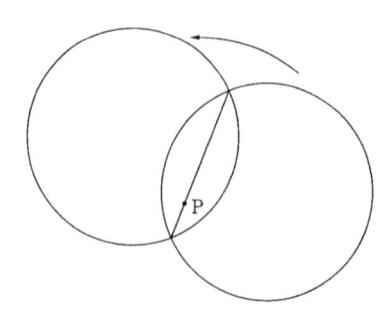

Figure 13: Four Fixed Points on an Intersecting Line of Two Circumscribing Circles

Figure 14: Matcing Circles

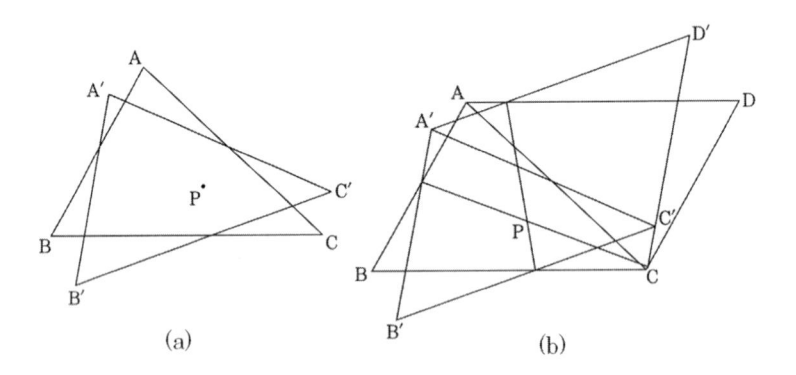

Figure 15: Fixed Point of Triangles

1/20,000 scale map in the same position as the previous two maps. The limit of such a process that is repeated infinitely, according to Coxeter, becomes a fixed point. This is a special case among those in which Brouwer's fixed point theorem, "an arbitrary continuous mapping has at least one fixed point," applies.

Knowing the existence of a fixed point does not necessarily mean knowing a method to construct it. As demonstrated by congruent transformation, Nishiyama's theorem is extended to construct a fixed point for similar transformation. Figure 17 shows Rectangle ABCD and Rectangle A'B'C'D' above, a rectangle scaled-down from Rectangle ABCD. Points P, Q, R and S are the intersections of sides of Rectangle ABCD and corresponding sides of Rectangle A'B'C'D'. When the sides of the two rectangles do not intersect, the sides are extended to make an intersection. Point O, the intersection of two diagonal lines PR and QS, is a fixed point.

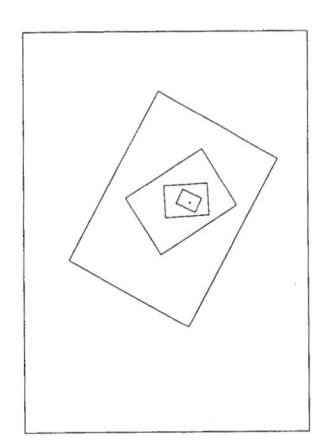

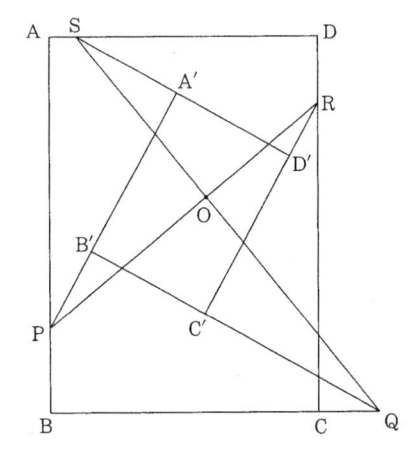

Figure 16: Fixed Point of Similar Transformation

Figure 17: Construction of a Fixed Point

Norio Morihara provides the following proof that Intersection O is a fixed point [5]. A horizontal line on Rectangle ABCD below is moved from BC to AD. A corresponding horizontal line on Rectangle A'B'C'D' above is moved from B'C' to A'D'. Line QS can be regarded as the locus of intersections between the horizontal line on Rectangle ABCD and the corresponding horizontal line on Rectangle A'B'C'D' moving at the same time. Similarly, Line PR can be regarded as the locus of intersections between a vertical line on Rectangle ABCD moving from AB to DC

38   CHAPTER 3

and a corresponding vertical line on Rectangle A'B'C'D' moving from A'B' to D'C'. A fixed point must meet conditions both vertically and horizontally. Therefore, the fixed point is Point O at the intersection of Line PR and Line QS.

   This mathematical demonstration by Morihara can be also applied to a fixed point of congruent transformation explained in Section 2 above. A fixed point of similar transformation is also confirmed by a random dot pattern. When a random dot pattern is microcopied onto an OHP film, the reduction ratio should be up to 90%. A reduction ratio of 50% would make confirmation of the fixed point difficult. As seen in Figure 18(a), only one fixed point surfaces, but the pattern surrounding it is different from that in congruent transformation. When the film above is rotated slightly, a spiral, not a concentric circle, appears. Right-handed rotation creates a dextrorsal spiral, and left-handed one brings a sinistrorsal spiral. When the film is moved in parallel, a radiological pattern is created around a fixed point as shown in Figure 18(b).

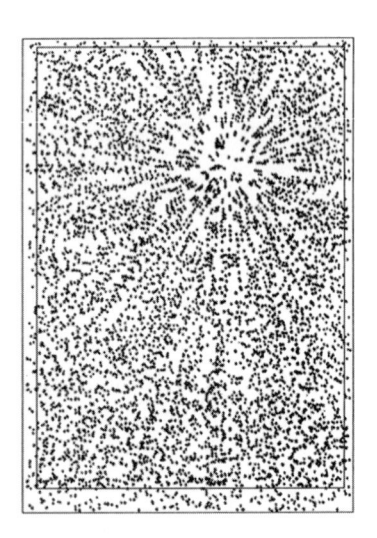

(a) Spiral pattern created   (b) Radiological pattern created

Figure 18: Random Dot Pattern for Similar Transformation

6. Computer Program to Produce a Random Dot Pattern

Since many people now have access to a computer, let me introduce a computer program using Visual BASIC to produce a random dot pattern. With such a program, a random dot pattern is easily produced. A built-in RND function is used to generate random numbers. The RND function automatically calculates uniform random numbers in $(0, 1)$ intervals. With the size of a square fixed, multiplication by random numbers decides the (x, y) coordinates. I assume here that each side of the square is 8000, as indicated by $wx = 8000$ and $wy = 8000$. Larger dots make for easier plotting, and "+" is formed by Line command. The number of dots is 2000. Only 14 lines constitute the command.

```
Private Sub Command1_Click()
wx = 8000
wy = 8000
sx = 100
sy = 100
Line (sx, sy)-(wx + sx, wy + sy), , B
For i = 1 To 2000
X = Rnd * wx + sx
Y = Rnd * wy + sy
d = 30
Line (X - d, Y)-(X + d, Y)
Line (X, Y - d)-(X, Y + d)
Next i
End Sub
```

Table 1. Visual BASIC Program

A random dot pattern appears on the screen when you run the program. The screen is then hardcopied. If the copy is not a square, you should adjust wx or wy. The pattern created is then printed on OHP film. A copy machine can be also used as printing equipment. Please try this process to make a set of materials for the random dot pattern yourself.

I would like to introduce another mysterious phenomena caused by a random dot pattern, which is used to demonstrate a fixed point. As shown in Figure 19, two random dot patterns are overlaid, and the OHP film above is rotated slightly to create a concentric circle in the center. When the film is moved horizontally, the concentric circle makes a vertical movement. When the film is moved vertically, it makes a horizontal movement. The direction of the film's movements and that of the concentric circle's movements, or the fixed point's movements, have a near-orthogonal relationship. Although an angle formed by these two movements is not an exact right angle, it infinitely approaches 90 degrees as the rotational angle goes toward zero. This relationship is similar to precession, which can be seen in the orthogonal movement of a spinning top's axis when you touch the axis with your fingertip.

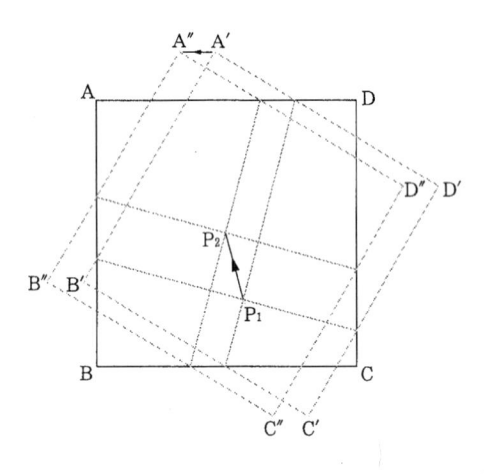

Figure 19: Movements of the Film and the Concentric Circle

The moving velocity of a concentric circle presents another interesting phenomenon. Only a slight movement of a film brings about rapid movement of the concentric circle. The velocity ratio between the film and circle is as much as several dozen times. The reason for such dynamic movement can be clarified by analyzing Figure 5 for fixed point construction. I encourage you to observe this for yourself.

References

[1] H. S. M. Coxeter, *Introduction to Geometry*, John Wiley and Sons Inc., (1965).

[2] Y. Nishiyama, Origami wo Soroeru [Coordinating Origami], *Sugaku Semina* [*Mathematics Seminar*], 21(2), (Feb. 1982), Cover, 28.

[3] Y. Nishiyama, Fudoten wo Omise Shimasu [Demonstrating a Fixed Point], *Sugaku Semina* [*Mathematics Seminar*], 21(8), (Aug. 1982), Cover, 122.

[4] Y. Nishiyama, En wo Kasaneru [Overlaying Circles], *Sugaku Semina* [*Mathematics Seminar*], 25(11), (1986), Cover, 67-69.

[5] T. Okabe, Ereganto na Kaito wo Motomu [Elegant Answers are Required], *Sugaku Semina* [*Mathematics Seminar*], 28(4), (1989), 87-89.

[6] J. Walker, The Amateur Scientist: Visual illusions in random dot patterns and television "snow", *Scientific American*, 242(4), (1980), 136-140.

CHAPTER **4**
Stairway Light Switches

Abstract: This article explains how stairway light switches can be turned on or off from different places. This idea is based on the binary system of Boolean algebra. Mathematics is hidden everywhere in daily life and we may enjoy it if we study the mathematics relating to everyday things.

AMS Subject Classification: 03G05, 00A09, 97A20
Key Words: Switching theory, Boolean algebra, Truth table, Venn diagram, Binary system

1. An Issue at Close Hand

Mathematics clearly serves many purposes, even though many people do not realize it. A good example in daily life involves the use of stairway light switches. Figure 1 shows a model of a stairway light switch of my own design. Before climbing the stairs at night, you first turn on the landing light using the switch in the hallway. Then, after reaching the top, you turn the light off using the switch on the upstairs landing. Everyone must have used such switches to turn the same light on or off from different places. This is an excellent application of mathematics, using a device conceived from the principle of binary numbers.

There are very many examples of mathematics being applied all around us. While we can't pay attention to every application, the switch on the stairs is certainly a useful example. And as mathematics wasn't invented to make students suffer in exams, I like to use such simple examples to help us understand that mathematics actually makes our lives easier.

So, what is it about this switch on the stairs? Remote controlled devices are very popular, so is it remote controlled? No, it's not that. There are also sensors that automatically activate when it becomes dark, but it's not that either. Are the switches on the different floors perhaps

Figure 1: Stairway Switches on Two Different Floors

connected, like the "bedside light switches" known to my generation that allow us when we tire of reading comic books and want to go to sleep to turn off the beside lamp using a cord, without getting out of bed? This type of switch has been very successful. But that's also not it. The answer in fact concerns the *relationship* between the electric wiring and the switch, and the problem begs for the kind of elegant solution a mathematics enthusiast would produce.

The switch on the first floor has two states. The switch on the second floor also has two states. These two states are lit (on) and unlit (off), and may be thought of as being related like binary numbers. So, let's begin by thinking about binary arithmetic problems.

2. Binary and Hexadecimal

Problem 1. Express the decimal number 61 in binary and hexadecimal.

Divide 61 by 2 and obtain the quotient and remainderof 30 and 1 respectively. Dividing 30 by 2 yields a quotient of 15 and a remainder of 0. These operations are repeated until a quotient of 1 is reached that cannot be further divided, and the steps of the calculation are recorded as shown in the diagram above. Next, the numbers 11101 written in bold are read off, beginning with the last quotient of 1 and proceeding upwards as indicated by the arrow. 111101 is thus the binary expression.

```
2 ) 61    Remainder
2 ) 30       1
2 ) 15       0
2 )  7       1
2 )  3       1
     1       1
```

Considering hexadecimal numbers in terms of binary, they take 4 digit blocks (4 bits). This is because $2^4 = 16$. The conversion to hexadecimal is therefore made simple by splitting 111101 into 4 digit blocks starting from the lowest position. Splitting it into 11 and 1101, and using the fact that 11 is 3, and 1101 is D, the hexadecimal expression can be found, which is 3D.

The origin of the term binary (or base 2) lies in the fact that there are two symbols, '0' and '1', used to represent numbers. These symbols are known as numerals. Being called base 2, we might expect the symbol '2' to be included, but counting '0' and using '1' means that there are already two symbols, so '2' is not included in base 2.

Decimal (or base 10) uses the ten numerals from '0' to '9', and express the values 0 to 9. In the case of hexadecimal (or base 16), there are sixteen symbols used to express numbers. The numerals used to represent 10 and above are the alphabetical characters 'A' to 'F'. The decimal value of 10 is represented by 'A', 11 by 'B', 12 by 'C', 13 by 'D', 14 by 'E' and 15 by 'F'. The decimal value of 16 is then written 10 ('one zero') in hexadecimal.

Numbers such as 11 ('eleven') in decimal and 11 in binary are easy to confuse, so in the binary case it's common to pronounce 1 as 'one', 0 as 'zero', and 11 as 'one one'. A subscript of 2, 10 or 16 may also be written on the right hand side to clarify whether a number is written in binary, decimal or hexadecimal, respectively. The answer to Problem 1 is written as follows.

$$61_{10} = 111101_2 = 3D_{16}$$

After obtaining the answer, it's also important to check back to see if it is correct. The way to convert the obtained binary number back into decimal is as follows.

$$111101_2$$
$$= 1 \times 2^5 + 1 \times 2^4 + 1 \times 2^3 + 1 \times 2^2 + 0 \times 2^1 + 1 \times 2^0 = 32 + 16 + 8 + 4 + 0 + 1$$

$$= 61_{10}$$

Problem 2. Calculate the sum, difference, product and quotient of the binary numbers $A = 1011$ and $B = 111$.

The four arithmetic operations in binary are as follows.

$$
\begin{array}{cccc}
1011 & 1011 & 1011 & 1 \\
+111 & -111 & \times111 & 111\overline{)1011} \\
\hline
10010 & 100 & 1011 & 111 \\
& & 1011 & \overline{100} \\
& & 1011 & \\
& & \overline{1001101} & \\
\end{array}
$$

$$\text{Addition} \qquad \text{Subtraction} \qquad \text{Multiplication} \qquad \text{Division}$$

In decimal, these are the four arithmetical calculations for the numbers $A = 11$ and $B = 7$. Check each one!

3. Boolean Algebra and Truth Tables

Problem 3. Think about the mechanism of the stairway switch that can turn the light on or off on two different floors. How does this mechanism involve maths?

Normal switches are known as single state switches where in the "on" position the light would be on, and in the "off" position the light would be off. However, the on and off positions of the stairway switch cannot be distinguished. It is the *relationship* between the switches on the first and second floors that determines whether the light is on or off.

Let us denote the switch on the first floor as A, its two states as 0 and 1, and likewise denote the switch on the second floor as B and its two states as 0 and 1. When the light is on its state is taken to be 1, and when it is off its state is 0. The switches A and B and the light each have only two states, so this is clearly a binary world.

The relationship between switches A and B and the light is summarized in Table 1. This kind of table is known as a truth table. There are two cases for switch A, and two for switch B, so in total there are $2 \times 2 = 4$ cases.

	(1)	(2)	(3)	(4)
A	0	1	1	0
B	0	0	1	1
The light	1	0	1	0

Table 1. Truth Table (Switches in Two Different Places)

This may also be expressed as a Venn diagram, as shown in Figure 2. The cases when the light is on are shown in the diagonal region. Venn diagrams are a special case of Euler diagrams. The region shown in this Venn diagram represents the cases when switches A and B are both 0, or when A and B are both 1, and in these cases the light is on. In the truth table these are cases (1) and (3).

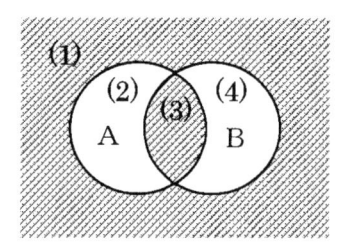

Figure 2: Venn Diagram (Switches in Two Different Places)

Writing this as a logical formula yields the following.

$$shining = AB + \bar{A}\,\bar{B}$$

AB is the product of A and B, $+$ indicates a sum, $\bar{A}$ is the negation of A and $\bar{B}$ the negation of B. A circuit of this type is known as an equivalence circuit. An equivalence circuit is the logical negation of exclusive or (Exclusive OR, XOR) as shown below.

Computers are constructed using binary as a basis. Electronic circuits are constructed from the following six fundamental circuits: logical and (AND), logical or (OR), negation (NOT), negative logical and (NAND), negative logical or (NOR) and exclusive logical or (XOR).

There are four basic arithmetic operations (addition, subtraction, multiplication and division). Multiplications are repeated additions, and divisions are repeated subtractions. All of the four arithmetical operations can therefore be expressed using the fundamental operation of addition in the one digit case.

48   CHAPTER 4

When thinking about such single digit addition, two locations for the result are necessary, the 'sum' and the 'carry'. For the sum, the following four patterns must be satisfied; the sum of 0 and 0 is 0, for 0 and 1 it is 1, for 1 and 0 it is 1, and for 1 and 1 it is 0. Likewise, the carry is 0 for 0 and 0, 0 for 1 and 0, 0 for 0 and 1, and for 1 and 1, it is 1.

$$\begin{array}{cccc} 0 & 0 & 1 & 1 \\ +0 & +1 & +0 & +1 \\ \hline 0 & 1 & 1 & 10 \end{array}$$

The addition circuit that provides these both simultaneously only requires an exclusive logical or (XOR) circuit for the 'sum' location, and a logical and (AND) circuit for the 'carry' location. Make a Venn diagram for each to confirm this. Now, I mentioned that the exclusive logical or and the stairway switch circuits are related by negation, but let's confirm this using logic equations. The exclusive logical or of A and B is written as $A \oplus B$. The fact that exclusive logical or is the negation of the equivalence circuit may be confirmed using equations by means of De Morgan's law, and is also made clear by Venn diagrams.

$$\overline{AB + \bar{A}\bar{B}} = \overline{AB} \times \overline{\bar{A}\bar{B}} = (\bar{A} + \bar{B})(\bar{\bar{A}} + \bar{\bar{B}})$$
$$= (\bar{A} + \bar{B})(A + B) = \bar{A}A + \bar{A}B + \bar{B}A + \bar{B}B$$
$$= A\bar{B} + \bar{A}B$$

By replacing binary calculations with logical set operations, the above equations can be reformulated using Boolean algebra. Boolean algebra is the basis of computing today. George Boole (1854), who formulated the system, made a great contribution without yet experiencing the emergence of computers. Some people say that mathematics anticipates developments 100 years in the future, but perhaps this is a romantic notion that only mathematicians can appreciate.

Now, back to the stairway switch. Maybe you have realized that a normal single state switch is no good for this problem. In fact the solution uses something called a 3-way switch (Figure 3). Since the light is only on when switches A and B are both 0 or both 1, *i.e.*, when they both have the same value, there is not just one wire between the switches on the two floors, but two. The final circuit diagram is as shown in Figure 4.

Off (state 0)   On (state 1)
(a) Single state switch

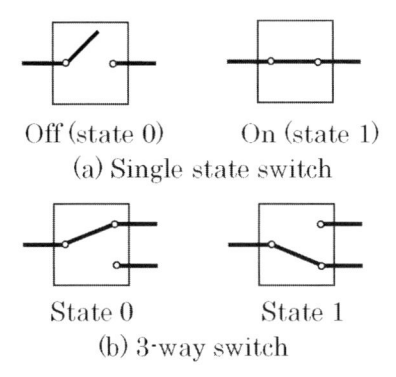

State 0   State 1
(b) 3-way switch

Figure 3: Single State Switch and 3-Way Switch

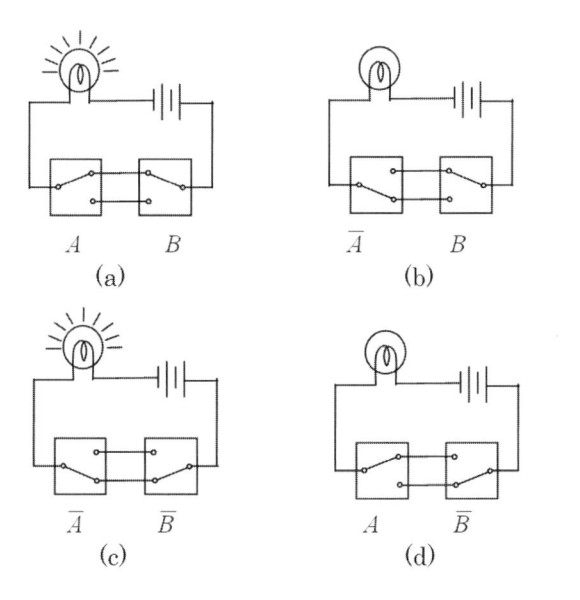

A B
(a)

$\overline{A}$ B
(b)

$\overline{A}$ $\overline{B}$
(c)

A $\overline{B}$
(d)

Figure 4: Switches in Two Places (Set of Four)

4. An Infinite Story House

Problem 4. Consider the design of a light switch system that can turn the same light on or off in three different places.

Now, the switch system in our house that can be used in the hall and on the upstairs landing to turn on and off the same light is extremely useful. This type of design can be used not only for switches in two locations, but also in three locations, or five locations, and such devices are in fact available in practice. Let's think about the three-location switch system. Figure 5 shows a switch system I designed for a three story house.

Figure 5: Stairway Switches on Three Floors

The three switches are denoted A, B and C, and the relationship between them determines the state of the light. Since each of the switches has the states 0 and 1, there are $2 \times 2 \times 2 = 8$ different cases. For each of these cases, denoting the situation when the light is on by 1, and off by 0, we can draw up a truth table as shown in Table 2.

	(1)	(2)	(3)	(4)	(5)	(6)	(7)	(8)
A	1	1	1	1	0	0	0	0
B	1	1	0	0	1	1	0	0
C	1	0	1	0	1	0	1	0
The light	0	1	1	0	1	0	0	1

Table 2. Truth Table (Switches in Three Places)

Expressing this as a logical formula yields the following.

$$Shining = AB\bar{C} + A\bar{B}C + \bar{A}BC + \bar{A}\bar{B}\bar{C}$$

It's necessary to think about the corresponding circuit, but rather than being composed of electric circuits for AND, XOR, *etc.*, it is actually simple to construct.

The cases when the light is on are (2), (3), (5) and (8). The cases when it is off are (1), (4), (6) and (7). Comparing these reveals that it is on when the sum $A + B + C$ is an even number, and off when it is odd. If instead the light were on when $A + B + C$ was odd, the functionality would be the same.

Now then, time for the actual circuit. Firstly, the switches on the first and third floors, A and C, use the 3-way switch shown above. The switch on the second floor uses one called a 4-way switch. The 4-way switch takes a value of 1 in the parallel condition, and 0 in the cross condition. The functionality of the cross condition is complicated, and its realization is shown in Figure 6(b).

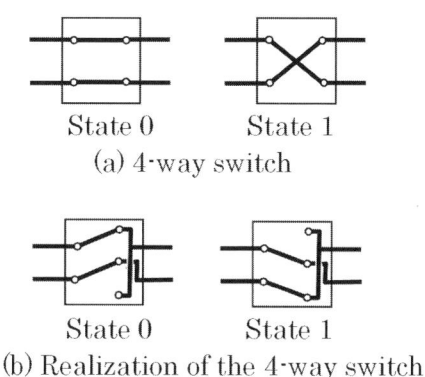

State 0   State 1
(a) 4-way switch

State 0   State 1
(b) Realization of the 4-way switch

Figure 6: Four-way switch

Let's confirm that the three location switch operates correctly by means of Figure 7. Suppose that the light is initially off. Switches A, B and C are in states 1, 0 and 0 (case a). Next, the light is turned on with switch A. Switches A, B and C are now in states 0, 0 and 0 (case b). Next, the light is turned off using switch B. Switches A, B and C are now in states 0, 1 and 0 (case c). Next, switch C is pressed to turn on the light. Switches A, B and C are now in states 0, 1 and 1 (case d). Then the light is turned off using switch B. Switches A, B and C are now in states 0, 0 and 1 (case e).

The thinking behind the three location switch can be applied to an unlimited number of locations to build our infinite story house. By using a 3-way switch at the left and right ends, and 4-way switches for all the switches in between, a circuit can be made by which a light can be turned on or off using switches in ten locations or even a hundred locations. The idea behind this kind of circuit is a by-product of computer technology, and more details can be found in [1].

Reference

[1] Y. Nishiyama, Byodona Switches [Equivalence Switches], in *Tamagowa Naze Tamago Kataka* [*Why are Eggs Egg-shaped?*], Tokyo: Nihon Hyoronsha, (1986), 127-144.

LIGHT SWITCHES 53

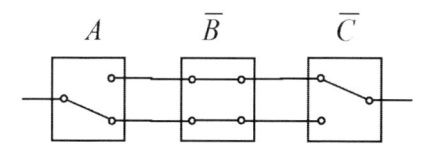

(a) Initially the light is off

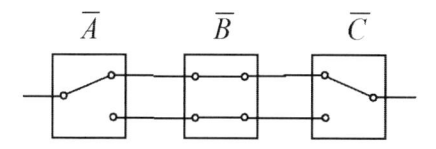

(b) Turned on with switch A

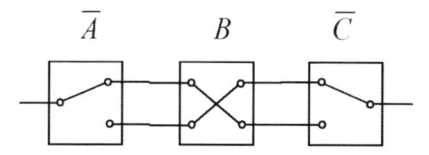

(c) Turned off with switch B

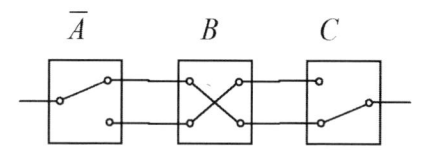

(d) Turned on with switch C

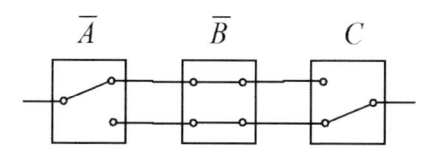

(e) Turned off with switch B

Figure 7: Confirming the Three Location Switch

CHAPTER 5
The Mathematics of Fans

Abstract: This paper explains why electric some fans appear to rotate backwards because of a stroboscopic effect. The phenomenon is often seen with fans in movies. After estimating fan rotation by mathematical formulae, many applications of stroboscopes are shown.

AMS Subject Classification: 00A08, 00A09, 97A20
Key Words: Stroboscopic effect, Virtual rotation, Alternating current, Movie frames

1. Some Fans Appear to Rotate Backwards

Electric fans are no longer seen so often, due to the popularity of air conditioners. Even so, I receive questions about fans from time to time, like "Some fans appear to rotate backwards. Why is this so?"

About 30 years ago, I wrote an essay entitled "Mathematics behind fans" that suggested that if we can understand the reason for the apparent backward rotation of fans, then we could use that fundamental principle to estimate the speed of a fan's rotation [1]. Let me explain the reason for that by supplementing what I wrote then to account for more modern environments.

The number of fan blades is generally either 3 or 4. Let's assume that it is 3, and that when the fan is powered on its motor begins to rotate in a clockwise direction. As the speed of rotation increases, however, one can see rotations that differ from the actual rotation. These could be clockwise rotations slower than the actual speed, a stationary appearance, or counterclockwise rotations in the opposite direction to the motor's rotation. These three states are possible. They are referred to as virtual rotations, which can be forwards, stationary or backwards (see Figure 1).

56   CHAPTER 5

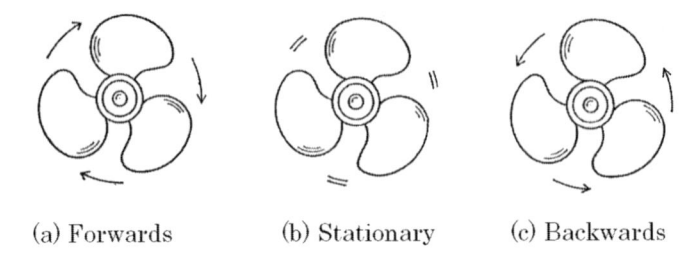

(a) Forwards   (b) Stationary   (c) Backwards

Figure 1: Three Virtual Rotations

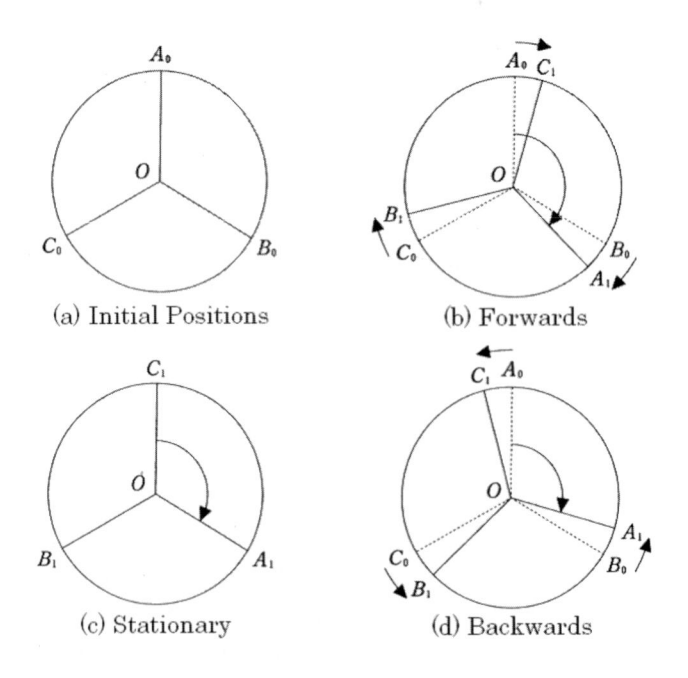

(a) Initial Positions   (b) Forwards

(c) Stationary   (d) Backwards

Figure 2: Actual and Virtual Rotations

All of these can be explained by the term "stroboscopic effect." Strobe light is a shortening of "stroboscopic light," a flashing lamp used for photography. The rotating visual phenomenon is possible when fans are observed under intermittent fluorescent or candescent lights. Such virtual rotations cannot be observed if the fans are taken outdoors and are observed under the continuous light of the sun, although I may be the only one who has conducted such an odd experiment.

Fluorescent and candescent lights are powered by an alternating current (AC). A direct current (DC) is constant and DC lights are also constant. Alternating current (AC) on the other hand, is not constant. AC has a wave like a Sinusoid with either 50 or 60 cycles per second in Japan. Thus, under such intermittent AC power, a light is also intermittent. Humans are known to recognize only about 10 movie frames per second. This is why we do not sense that the light is blinking intermittently.

So what happens when we observe the fans under intermittent light? Let's illustrate the answer using Figure 2:

Let's denote the three blades as A, B and C. Let us further assume that the initial blade positions are OA_0, OB_0 and OC_0, respectively (Figure 2(a)). Assume also that, when the light comes on after one AC cycle, the blades have advanced by about 120 degrees to the new positions OA_1, OB_1 and OC_1, respectively. At this time, our eyes do not realize that blades have moved from OA_0 to OA_1, OB_0 to OB_1 and OC_0 to OC_1. Instead, we erroneously regard the blade at the position closest to the original blade's previous position as the original blade. In other words, we feel that OA_0 has moved to OC_1, OB_0 has moved to OA_1 and OC_0 has moved to OB_1.

Therefore, during one cycle in which the light blinks, the fans move forward virtually if the fans advance by slightly more then 120 degrees. The fans are virtually stationary if the fans advance exactly by 120 degrees, and the fans move backwards virtually if the fans advance by less than 120 degrees (Figure 2(b) - (d)).

2. Alternating Current and Rectification

What then is the frequency of the blinking fluorescent and candescent lights? As you may know, the frequency of alternating current in Japan is 60 Hz in Western Japan and 50 Hz in Eastern Japan.

Why does a small country like Japan have two different frequencies? During the Meiji Restoration, there were lingering power conflicts among political sections, each of which purchased power generators from foreign countries with close ties. Tokyo Dento (Tokyo Electric Power today) adopted a 50 Hz AC power generator from Allgemeine of Germany in 1895, while Osaka Dento adopted a 60 Hz AC power generator from GE of USA in 1897. Thus, 50 Hz was adopted as the standard for Eastern Japan, while 60 Hz became the standard for Western Japan. Although there have been several opportunities to unify the two frequencies, these two have remained separate until today.

My initial understanding was that 50 Hz or 60 Hz AC causes lights to blink 50 or 60 times per second, but some of my readers wrote to me to inform me that because of a process called AC rectification, the lights actually blink twice as fast, *i.e.*, 100 or 120 times.

This is illustrated in Figure 3. Let's assume we are using 50 Hz AC power. Unlike DC, AC is represented by a Sine wave. The wave has positive and negative regions, of which only the positive area is significant for the current. This section constitutes a half-wave rectification. In this state, the blinking occurs at 50 Hz. Some books explain that this half-wave rectification is sufficient for lights not requiring much illumination, like a tail lamp of a bicycle. This half-rectification was possibly used immediately after the war. When two rectifiers are used, converting the current direction of the negative region, a full-wave rectification is created where the light blinks at 100 Hz. Most of the lights today use this full-wave rectification.

In this manner, the lights blink at 100 Hz or 120 Hz. Thus, it is difficult to see virtual rotations if the fans are observed under electric lamps as explained above. Let me explain how the virtual rotations can in fact be observed. In order to do so, I must first explain the fan's specifications.

I once wrote a letter to an electric appliance company asking for the rotational speeds of their fans for "research purposes." I believe the rotational speeds have not changed so much since then. One fan for 60 Hz Western Japan had three settings. They were a light breeze at 720 revolutions/minute, a cool breeze at 1040 revolutions/minute and a strong breeze at 1390 revolutions/minute.

Although the motor revolutions are often expressed per minute, the electric light frequency is usually expressed per second. Let me therefore convert these numbers to a scale of seconds. The light breeze is 12 revolutions/second, the cool breeze is 17.3 revolutions/second and the strong breeze is 23.2 revolutions/second. The average comes to roughly 20 revolutions/second. This means that if the fan has three blades, under an electric light at 60 Hz, it rotates around 120 degrees every

(a) Electric current (50 Hz)

(b) Half-wave rectification (50 Hz)

(c) Full-wave rectification (100 Hz)

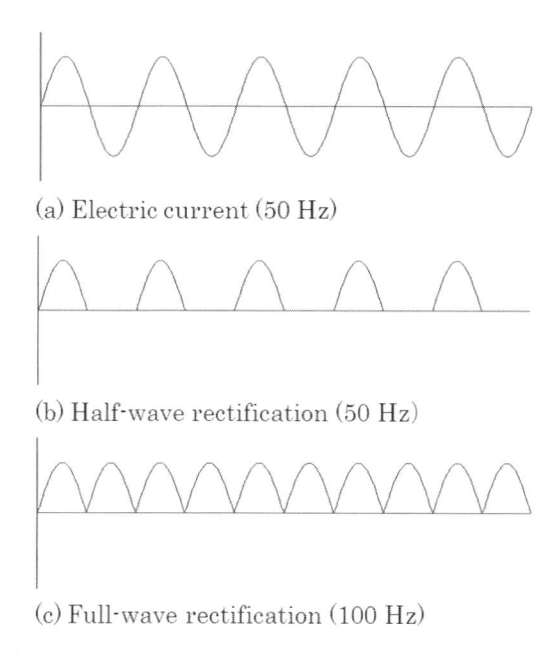

Figure 3: Rectification

1/60th of a second. In other words, before and after the fan goes into the strong breeze speed, we should be able to observe forward, stationary and backward virtual rotations.

	50Hz	60Hz
Light breeze	800(13.3)	720(12.0)
Cool breeze	1050(17.5)	1040(17.3)
Strong breeze	1280(21.3)	1390(23.2)

Unit: Rotations/minute, numbers in parentheses show rotations/second.

Table. Fan Rotation Speeds

However, it is difficult to observe these phenomena because the electric light frequency is either 100 Hz or 120 Hz due to rectification as explained above.

What should we do then? Give up the fluorescent light and candescent light in favor of TVs. Braun tube TV scanning lines transmit 30 image frames per second. 30 Hz is a smaller frequency than that of the fluorescent lamps. When you wave your hands in front of a TV display, your hands seem to flicker. This is because of the intermittent images displayed 30 times per second by the TV.

TVs using Braun tubes use 525 scanning lines which run from the top left corner to the bottom right corner 30 times per second. However, since this TV image flickers and is not good for human eyes, an interlace scanning method which scans odd number lines at 60 Hz and even number lines at 60 Hz has been used more lately. Furthermore, since TVs have moved from old Braun tubes to LCD and plasma displays, it might not be possible to see such phenomena these days. So, it is true that there are fewer opportunities to observe fans rotating backwards these days.

3. Fans in the Movies

However, I recently received the following question. Apparently some fans are seen rotating in a backwards direction, not under an electric light, but in the daylight. This suggests that the stroboscopic effect is not the reason for the backwards rotation.

We can imagine two scenarios here. Were these fans rotating in a backwards direction observed under actual sunlight, or were they seen on TV screens or in movies? In the latter scenario, since the movie films are not displayed continuously but are made up from a certain number of still images shown each second, a stroboscopic effect would result. Movies show 24 frames per second, while TVs show 30 frames per second.

There is a history behind this frequency of 24 frames per second in the movies. The projection speed of the movies was obtained empirically based on afterimages, which are a human visual characteristic. Initially, it was 10 frames per second. But this did not produce smooth images, and it was increased to 16 frames per second. The reason for the low number of frames was to minimize the consumption of film itself, which was very costly in those days. Furthermore, the film was flammable and higher speeds with more frames per second also increased the risk of fire. With the advent of talkies, when sound was recorded on the edge of the film, 16 frames per second produced jerky sounds. The speed was thus eventually increased to 24 frames per second, which is the world standard today.

What would happen to the images if fans were recorded with a 24-frame movie or by a 30-frame TV? As I explained above in relation to stroboscopic effects of fluorescent lamps and incandescent lamps, when we observe images at 24 or 30 frames per second, the same phenomenon occurs. Since movies and TVs use smaller frequencies than electric lights, it is easier to observe the virtual rotations. Many of us have seen fans in old black-and-white movies, wheels of covered wagons in Westerns, helicopter blades, and automobile wheels in TV commercials all rotating backwards.

4. Estimating Fan Rotations

Let me try to explain this effect using mathematical formulae. Let the strobe light frequency be n_1 (Hz) and the fan frequency be n_2 (Hz). Both of these are periodic functions which can be expressed as Sine waves. Let the time be $t(0 \leq t \leq 1)$. Then we get:

$$y_1(t) = \sin 2\pi n_1 t, \quad (1)$$

$$y_2(t) = \sin 2\pi n_2 t. \quad (2)$$

At this time, the frequency of the observed virtual rotation n_3 (Hz) is related by $n_3 = n_2 - n_1$, which produces the following.

$$y_3(t) = \sin 2\pi(n_2 - n_1)t \quad (3)$$

As an example, if we observe a blade rotating at 9 rotations/second under a strobe light blinking at 10 rotations/second, we will see a virtual rotation at 1 rotation/second. In numerical values, this would be $n_1 = 10, n_2 = 9, n_3 = 9 - 10 = -1$. This is illustrated in Figure 4.

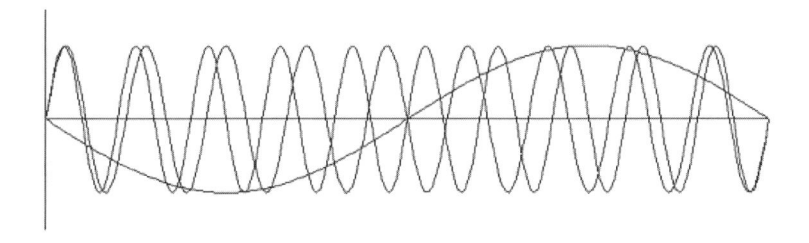

Figure 4: Relationship between a Rotating Object and a Strobe Light

Likewise, if a propeller rotating at a frequency of 100 Hz is illuminated under a 100 Hz light, the propeller should be observed as stationary. Furthermore, if 99 Hz is used for the illumination then the virtual rotation should move forwards slowly, in the same direction as the propeller at a speed of 1 Hz. A 101 Hz light would produce a slow virtual rotation moving backwards. Helicopter blades seem to rotate slowly in the movies if a multiple of the number of movie frames per second is almost the same as the number of revolutions of the propellers.

Let's place a fan in front of a TV screen and observe the fan. The TV receiver emits images at 30 Hz, *i.e.*, 30 intermittent images per second. If the blades rotate 1/3, 2/3, or 1 complete rotation after 1/30th second, the fan appears stationary. On a per second basis, such phenomena are observed when the blades are rotating $30/3 = 10$ times, $30 \times 2/3 = 20$ times and $30 \times 1 = 30$ times per second.

Let the actual rotation speed be n_1 rotations/second, and the virtual rotation speed be n_2 rotations/second. Then, we get the following relationship.

$$n_2 = f(n_1) = n_1 - 10i \quad (10i - 5 \leq n_1 < 10i + 5, \quad i = 0, 1, \cdots, m)$$

As the linear function on the left side of Figure 5 shows, rotations proportional to the rotational speed of the fan should be observed. However, as the saw-tooth function on the right side shows, virtual rotations are observed within a maximum and minimum range of ± 5 rotations/second. $n_2 < 0$ produces backwards rotation, $n_2 = 0$ produces a stationary appearance and $n_2 > 0$ produces forwards rotation. These are repeated.

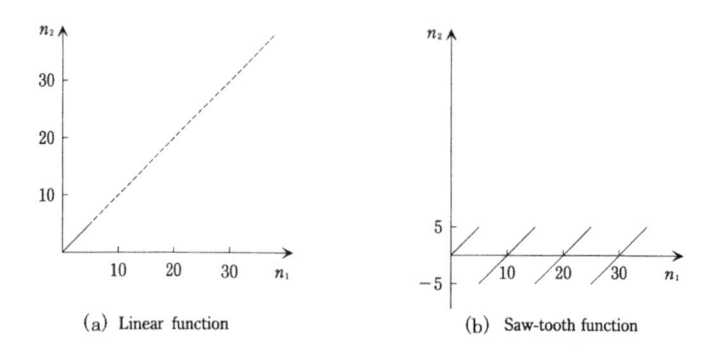

(a) Linear function

(b) Saw-tooth function

Figure 5: Numbers of Actual and Virtual Rotations.

Further, although the fan we were considering has only three blades, it is also possible to observe the blades in multiples of 3, such as 6 blades or 9 blades, depending on the location of the blades (Figure 6).

In fact, it is often possible to observe 12 thin blades. Armed with these observations, we can also estimate the actual rotational speed of the fan from the virtual rotations.

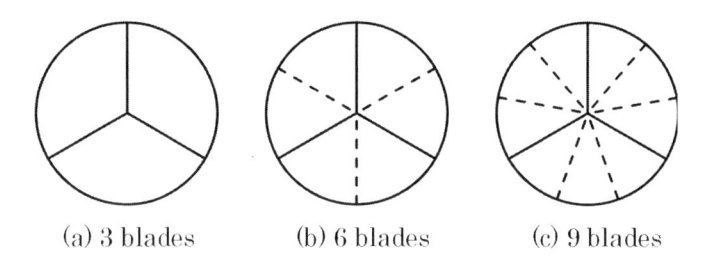

(a) 3 blades (b) 6 blades (c) 9 blades

Figure 6: Generation of Multiples of the Number of Blades.

5. Application of Stroboscopes

I have explained above that fans seem to rotate backwards due to stroboscopic effects. Stroboscopes make use of this basic principle. The following are some examples.

Figure 7 is an adjustment sheet for rotational speed, which was very popular during the heyday of analog players. The younger generation of today, at a time when CD players rule the world, may not be familiar with such devices. But in those days, the rotational speeds of players were 33 1/3 rotations/minute and 45 rotations/minute. Different models were available for the 50 Hz and 60 Hz areas. Unfortunately, I was unable to confirm this since I do not know what has happened to my own analog player, but under the old lamps, this pattern would have appeared stationary at the correct rotational speeds.

For industrial uses, there are non-contact speedometers. There were contact and non-contact speedometers, and the non-contact types utilized the stroboscopic effect.

The rotational speed of a revolving object could be measured without contact by attaching a reflective tape to the revolving object and illuminating it under a visible red LED light. According to the specification of one particular product, its accuracy is listed as ±0.02% for 1 - 99999 rotations/minute, which is fairly high. A 3V Xenon flash lamp is supposed to generate a strong intermittent light. Measurements are made by adjusting the flashing rate so that the subject of measurement appears to be stationary and then reading off the value indicating the

Figure 7: Rotational Speed Adjustment Sheet

rate at that moment. The subjects of measurement are revolving objects rotating at high speeds, such as motors.

Stroboscopes are also used with medical equipment. The vocal cords vibrate between 100 - 300 Hz for voicing. When there are disorders with the vocal cords, precise diagnosis during vocal emission requires accurate frequency measurements. Such examination of diseased sections is possible visually if a CCD camera or other imaging device is attached to an endoscope. There are also devices that observe molecules moving at super-high speeds using stroboscopes with laser lights. Stroboscopic effects are thus used for many purposes, but the basic principle behind all of them is the phenomenon which makes fans look as if they are rotating in a backwards direction.

Reference

[1] Y. Nishiyama, Senpukini Hisomu Suri [Mathematics behind fans], in *Tamagowa Naze Tamago Kataka* [*Why are Eggs Egg-shaped?*], Tokyo: Nihon Hyoronsha, (1986), 107-126.

CHAPTER 6
The Mathematics of Egg Shape

Abstract: This paper explains why the shapes of eggs are oval, and why eggs stop on slopes. After touching upon Descartes' and Cassini's oval curves, eggs are classified into 4 groups: oval, pyriform, circular and elliptical. Because only oval and pyriform eggs stop on slopes, it is explained that egg shape may be related to Darwin's theory of evolution.

AMS Subject Classification: 92B99, 00A09, 97A20
Key Words: Oval, Descartes' oval, Cassini's oval, Circle, Ellipse, Darwin's Theory of Evolution

1. Eggs Stop on Slopes

The main topic of interest in biology at the moment is DNA, but if you pay attention to shapes and numbers you will see that many mathematical elements are included, and in this essay I'd like to explain the peculiar shape of eggs. I made a presentation about this topic in the August 1979 issue of *Mathematical Seminar* under the title 'The shape of eggs,' which was a long time ago, but rereading it now, their shape seems just as strange [1].

Hasn't everyone thought that the shape of eggs used for cooking is strange? Eggs are oval, which means egg-shaped, so while the inquiry is just like a Zen riddle, let's reveal the secret little by little. In mathematics, a representative example of round forms is the circle. Circles may be determined by their center and radius. At high school we learn about ellipses, which are circles stretched sideways and their curve is determined as a constant sum of the distance to two foci (or fixed points).

Eggs are neither circular nor elliptical. Eggs are oval. If you observe an egg closely, the distance from the center is not a fixed circle. The horizontal aspect has a longer ellipse-like form. Observing closely once again, one horizontal direction is roundly curved but the other is pointed (Figure 1). This is the shape of an egg. The longer axis is called the major axis, and the shorter is called the minor axis. The rounded end is called the base and the pointed end is called the tip. Since eggs are actually three-dimensional bodies, they should not be expressed in terms of circles or ellipses but rather spheres and ellipsoids. However, it is sufficient to think in terms of cross sections, so I will explain here using planar shapes.

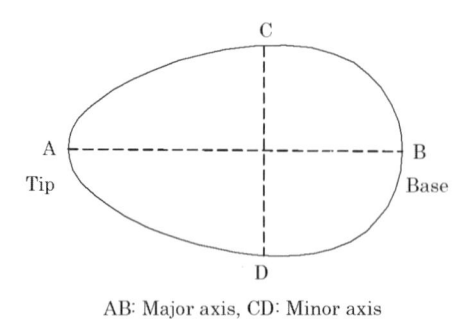

AB: Major axis, CD: Minor axis

Figure 1: The Shape of an Egg

Let's think about why eggs are shaped as they are. Place an egg on a tabletop. The major axis is not parallel to the surface of the table. The pointed end is closer to the surface, and it stabilizes with the rounded end further from the surface (Figure 2). For an ellipse the major axis would be parallel to the surface. If we consider the egg's center of mass we can understand why it stabilizes with the major axis tilted. For circles and ellipses, the center of mass is in the exact mid-point between the vertical and horizontal axes, but for an egg, since one end is pointed and the other is rounded, the center of mass is slightly offset towards the rounded end. What happens when an egg with its center of mass offset from the center is placed on a tabletop? As shown in Figure 2, the gravity force W from the egg's center of mass O, and the reaction force N from the contact point P lie on the same straight line, and the major axis stabilizes with a tilt. I explained this using the terminology of physics, but everyone knows that eggs tilt like this.

What happens because of the tilt in an egg's major axis? If an egg is placed on a slope, no matter what position it is placed in, it settles in a stable position without rolling away. This is peculiar. The stable position is with the pointed end oriented towards the top of the slope, and the rounded end towards the bottom of the slope (Figure 3). I'd like for those readers who have until now had no interest in the shape of eggs to begin by confirming this experimentally.

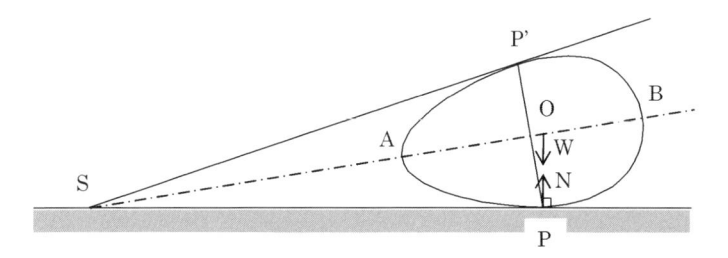

Figure 2: Eggs Tilt

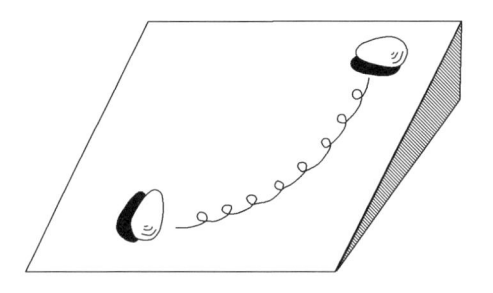

Figure 3: Eggs Settle on Slopes

My article "The shape of eggs" was subsequently included in an independent book, *Why Are Eggs Egg-shaped?* (Nihon Hyoronsha). I received only one protest regarding the article from a certain reader. An egg was placed on a tilted tabletop and rolled, but it did not stop. They proceeded to say that my claim was a lie, and ought to be revised. I became worried, and repeated the experiment, but the egg did stop after all.

So why didn't the reader's egg stop? Perhaps they angled the tabletop at about 30 degrees and then rolled the egg. The tabletop must be tilted less than 5 degrees, and it must be released carefully and gently. Figure 3 is a schematic diagram intended to show the slope, and if it is interpreted as showing a slope of 30 degrees it could cause a problem. Also, if the surface of the table is too smooth the egg might not stop. A certain degree of frictional resistance is necessary. In addition, I have also performed the experiment with a boiled egg, and sometimes it doesn't stop. A member of staff brought an egg to use for recording on a particular television program, but the egg was boiled. I was told that there is a greater risk of breakage during transportation with raw eggs, but the center of mass in the boiled egg is in a slightly different place and the egg did not stop. Also, when an egg is boiled it may cease to be rough and lose its friction. On that particular occasion the egg was exchanged for a raw egg at short notice.

It can be confirmed that paper cups stop on slopes in the same way as eggs when they are rolled. Paper cups usually have a large circular end to drink from, and a smaller circular base. If the side surface gripped by the fingers is extended, it forms a cone. Eggs and paper cups both approximate cones (when extrapolated). The problem of eggs or paper cups rolling on slopes can be replaced with the problem of a cone rolling on a slope. It's not that hard to think about why a cone stops rolling on a slope (Figure 4).

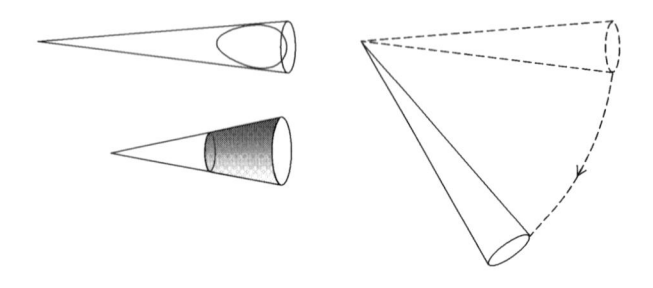

Figure 4: Eggs and Paper Cups Approximate Cones

Furthermore, eggs and paper cups also share the characteristic that, besides on slopes, they do not roll far on horizontal surfaces. Suppose that an egg rolls away when a bird is cradling eggs in order to hatch chicks. The bird cannot move to bring back the egg. In the same way that a cone describes a circular arc and returns to its original position, eggs also describe circular arcs and return to their parents.

2. Descartes and Cassini's Oval Curves

Descartes and Cassini's methods may be used to describe oval curves. In 1637, Descartes defined oval curves as follows. Two circles form the basis. One circle has center O_1 and radius r_1, while the other has its center O_2 offset in the x axis by a and has radius r_2. Two parallel lines are drawn, one going through each center, and the intersection point of each line with the other circle is denoted as either A or B (in this case O_1B and O_2A are parallel). The intersection point of O_1A and O_2B, P gives the coordinates of the oval. For $m, n > 0$, $m\overline{O_1P} + n\overline{O_2P}$ relates the parameters as a constant value. When $m = n$ the curve is an ellipse, and the radii of the two basis circles are equal ($r_1 = r_2$). Figure 5 was drawn using a Visual Basic program with the values $a = 1, r_1 = 1.2$ and $r_2 = 1.8$. It clearly has an egg shape.

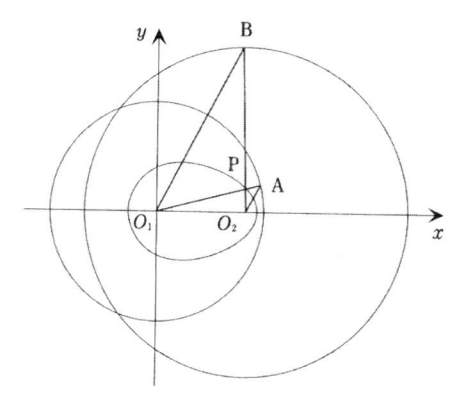

Figure 5: Descartes' Oval Curves

Another way to describe oval curves is Cassini's method. In 1680, Cassini defined the oval curve as follows. The trajectory of points X such that the product of the distances to two fixed points (or focii) is constant describes an oval curve (Figure 6). With O, the mid-point of A, B as the origin, and the line joining A and B as the x axis, the equations relating the orthogonal axes are as follows.

$$(x^2 + y^2)^2 - 2a^2(x^2 - y^2) = k^4 - a^4$$

Note that $\overline{AB} = 2a$ and $k^2 = \overline{AX} \times \overline{BX}$.

In particular, if $a^2 = k^2$ then O is the nodal point of the curve. In this case, the curve is known as a Lemniscate. Cassini's oval curve is expressed as an implicit function so it cannot be graphed in this form. If the substitutions $x = r\cos\theta$ and $y = r\sin\theta$ are made, then the function is represented in polar coordinates by the following equation.

$$r^2 = a^2 \cos 2\theta \pm \sqrt{a^4 \cos^2 2\theta + k^4 - a^4}$$

Contour drawing software can be used to draw the oval curve specified by the former equation. The latter equation can be drawn comparatively easily using a Visual Basic program.

In the case of an ellipse, the sum of the distance from the two fixed points is constant, but in Cassini's oval curve it is the product that is constant. Several ovals were drawn with different values for the parameter k. Figure 6 shows the diagrams for $k = 1.4, 1.2, 1$ and 0.98 when $a = 1$. When $k > a$ the outer curve is an elliptical oval. As k gets smaller the region near $x = 0$ gets thinner, and when $k = a$ it becomes the Lemniscate curve with origin O. For $k < a$ the curves divide into two and it becomes oval. The most interior curve is the egg shape we have been discussing here.

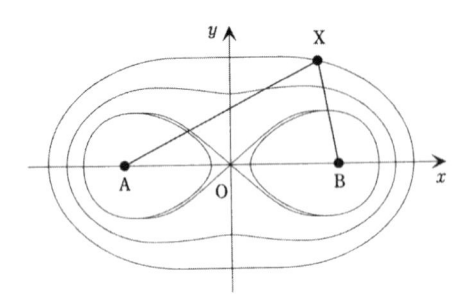

Figure 6: Cassini's Oval Curves

Cassini (1625-1712) was an Italian astronomer who was invited to Paris by Louis XIV and became the first director of the astronomical observatory. He made many accomplishments, including the measurement of Jupiter and Mars' periods of rotation, discovering the gap in Saturn's ring and its four moons, and calculating the distance between Mars and the Sun. Cassini thought the orbits of the planets were oval, but in fact Newton (1642-1727) showed that they are elliptical orbits with the Sun situated at their focal point. There are also some interesting properties such as with the torus (a doughnut or floating ring). If it is cut by a plane parallel to its axis of rotation, the cross section revealed is one of Cassini's ovals.

3. Eggs Have Various Shapes

In biology, 'egg' refers to an egg cell, but the word is usually used to mean the thing which is laid externally. The cytoplasm includes the nutritious material of the yolk, which is surrounded by various materials such as the white of the egg. There are various sizes of egg including the whale shark's at 68×40 cm, the ostrich's at 16×12 cm and a type of pigeon at 1.2×0.8 cm. (I was able to see these in 2005 when I had the opportunity for overseas study in Cambridge. I used a free day to visit the museum of natural history in London, where they had these eggs on display.) Paying attention to the shape of the eggs, they can be classified into 4 groups: oval, pyriform, circular and elliptical (Figure 7).

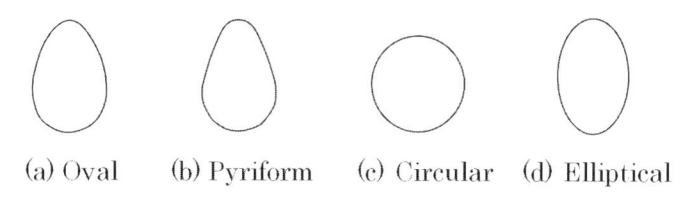

(a) Oval   (b) Pyriform   (c) Circular   (d) Elliptical

Figure 7: Types of Egg Shape

Chicken's eggs are representative of ovals. Pyriform eggs are common among seabirds such as Auks and murrelets, and the difference between the tip and the base is larger than for an oval. Round eggs are represented by sea turtles, and ellipses by ostriches.

Chickens now lay their eggs in flat places so there is no need to worry about them rolling away and breaking, and their eggs are oval. Perhaps the eggs are oval so that when the hen is brooding and keeping her eggs warm they don't roll far away but describe a circular arc and come back. The sea turtle lays its eggs in a sandy beach, and the ostrich lays them in grassland, both of which are flat places where there is no danger of the eggs rolling so it is understandable for the eggs to be round or elliptical. Auks and murrelets lay their eggs on narrow rocky shelves. It is common for the rock shelves to be sloped, so the reason behind their pyriform shape also makes sense. Gulls and gannets are also seabirds, but since their eggs are not pyriform there is a danger of them rolling, so they build nests. Thinking about it this way, one can understand that of course, that's the reason why eggs are egg-shaped!

4. Darwin's Evolutionary Theory

Were eggs always egg-shaped? Were the shapes of eggs fixed from the beginning, or were they set later?

Darwin proposed a theory of evolution known as the theory of natural selection. It was discovered in parallel by Darwin and Wallace at the same time, and is a theory about the cause of evolution. Proliferation is a principle of living things, and as a result of the consequent competition for survival, individuals with adaptations suited to their environments arise. These adaptations are transmitted to their descendants. The idea is that living things thus gradually advance in the direction of these adaptations to their environment and evolve. Darwin related this theory in full in *The Origin of Species*, and as a result the theory of evolution was widely acknowledged. Even in the 20th century it occupies a central place in the contemporary theory of evolution. I recommend casting an eye over *The Origin of the Species*.

As I will show below, egg shape may be seen as material supporting Darwin's theory of evolution. Figure 8 shows the structure of an egg.

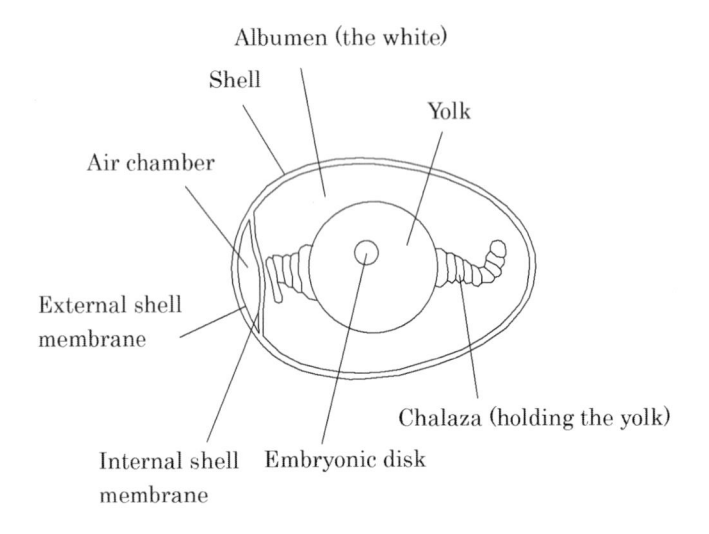

Figure 8: The Structure of an Egg (from the *Picture Book of Birds*, Shogakukan, p.160)

The construction of an egg includes its casing, the egg shell. Breaking this open reveals the albumen (commonly known as the white), the yolk, the chalaza which holds the yolk, the embryonic disk which is the origin of the bird's body, an air chamber, the shell membrane and so on. Surprisingly, while chicken's eggs are certainly oval, breaking one open and removing the yolk reveals that the yolk is close to spherical. The yolk is a ball. It is slightly distorted due to the influence of gravity, but it is basically round.

Do you know about *tamahimo*, which is sometimes sold by poulterers? It is chicken ovaries and oviducts (Figure 9). Besides eggs and chicken meat, these kinds of spoils are also put on sale. The *tama* in *tamahimo* refers to the ovaries, and the *himo* refers to the chicken's oviduct. While chickens that could no longer lay eggs were sold as meat, in the old days, the *tama* and *himo* were said to be nutritious and considered quite precious spoils.

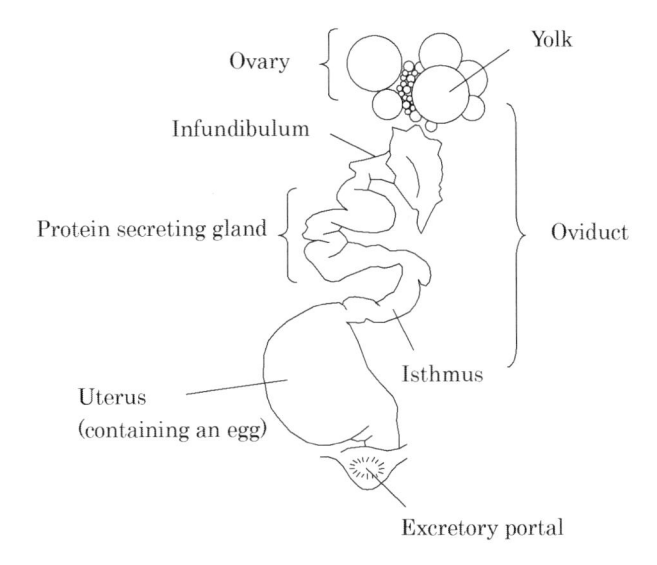

Figure 9: The Ovaries and Oviduct (from *Chicken and Egg*, M. Saito, Dowashunju)

Chickens lay one egg per day, but here we can examine the process behind the generation of a single egg before it is laid.

Inside the ovaries there are many yolks, like bunches of grapes. They all have an approximately spherical shape. When one is sufficiently large, it enters the oviduct via the infundibulum (funnel), and while receiving secreted proteins, the albumen and ephippium (also known as the soft shell) begin to form. The shell forms inside the uterus through the composition of calcareous material. The time between its passage into the oviduct via the infundibulum until laying is about 24 to 27 hours, and the total length of the oviduct is about 70 to 75 cm.

The shape of the shell is not determined at the moment of release from the ovary, but during the 19 to 20 hours that it is retained in the uterus. The formation of the shell in the uterus is related to dietary calcium. If there is a calcium deficiency then the shell will be thin. The yolks in the ovaries are spherical, but when excreted, the shell may be spherical, elliptical, oval or pyriform.

If an egg laid onto a rocky shelf were round or elliptical, then it would roll and break, so this type probably died out. If the egg were, even by chance, oval or pyriform then because of its shape it would not roll and the type would be saved. From this regard, while the distance from the ovaries to the excretory portal is only a few tens of centimeters, this path can be thought of as integrating the history of evolution since before archaeopteryx (in the Jurassic period).

5. Applications of Egg Shape

I have reached an explanation from a biological perspective that bird's eggs have the successful property of returning to their parents without rolling and breaking, but egg shapes are not limited to biology. There have been many attempts to use it in our daily lives. Searching for the keyword 'oval' on the internet reveals some interesting pages.

There are egg-shaped sludge digestion chambers. In comparison to previous tubular digestion chambers, it is explained that they are superior in terms of water- and air-tightness. There are egg-shaped gutters. The cross section of flowing water is egg-shaped, so they are said to have good drainage. There are egg-shaped car bodies. Streamlining has developed in cars with a front end corresponding to the tip of an egg and a rear end like the base of an egg.

There are egg-shaped mobile phones. In fact they are elliptical rather than oval, but they were also designed with eggs in mind. There are egg-shaped speakers. It is explained that by suppressing various oscillations and echoes occurring inside the speaker, the egg shape is used to pursue reproduction of the original sound. In addition, there are concept products such as egg-shaped refrigerators, egg-shaped washing machines and so on. That doesn't mean to say that they have all been scientifically proven, but it can be easily understood that the shape of eggs is not only relevant to biology, it also has a significant influence in our lives.

Reference

[1] Y. Nishiyama, *Tamagowa Naze Tamago Kataka* [*Why are Eggs Egg-shaped?*], Tokyo: Nihon Hyoronsha, (1986).

CHAPTER 7
What's in a Barcode? Duplicated Combinations

Abstract: This article explains how barcodes are read mathematically. One numeral is constructed from 7 modules in a barcode. Firstly, a counting method is presented through which the number of different characters that can be expressed is obtained. After touching upon formulae for permutations and combinations, duplicated combinations are expressed explicitly. The use of a modulus of 10 is mentioned with regard to the calculation of check digits.

AMS Subject Classification: 00A09, 90C08, 97A20
Key Words: Barcode, Tree diagram, Permutation, Combination, Duplicated permutation, Duplicated combination, Check digits, Modulo 10

1. How Are Barcodes Read?

I'm sure you know that commercial goods usually have a barcode marked on them somewhere. These barcodes include information related to maker and product codes, and they can in an instant be read-in and so reduce the burden on cashiers. Barcodes are also essential when operating the Point-of-Sale (POS) system [1].

We all know that barcodes are composed of a pattern of black and white stripes, but do you know what they mean? The truth is they have a close relationship with the combinations and permutations studied in counting patterns. Mathematics therefore has application to the manufacture of goods and its related technologies. Standard barcodes are composed of a total of 13 digits. From the leftmost digit, the first 2 represent the country code (Japan is 49), the next 5 are the maker code, the following 5 are the product code, and the last 1 is a check digit. In the example in Figure 1, the country code is 49, the maker code is 01306, the product code is 04282 and the check digit is 3 [2].

Let's try the following investigation to discover the relationship between a 13-digit barcode and its black and white striped pattern. Firstly,

Figure 1: An Example Barcode

the width of the lines varies, but how many lines are there in total? Careful counting reveals that there are 30. Assuming that these 30 lines represent 13 digits, there are 2.3 lines for each digit. However, doesn't such an inexact number raise a few questions? Examining the lines carefully, it can be seen that there is a sequence of long thin lines, with 2 at the left edge, 2 in the center and 2 at the right hand edge. These show where the barcode begins, its midpoint and where it ends, and do not represent actual numbers. Excluding these long thin lines leaves $30 - 2 \times 3 = 24$ lines.

The first digit, 4, of the country code 49 is not expressed within the barcode. Instead this 4 is printed just to the side. Compiling these facts, 24 lines represent 12 numerical digits, and since $24 \div 12 = 2$ so two lines represent each digit.

Next, let's look at the thickness of the lines. How many times wider is the thickest line compared to the thinnest line? Careful observation reveals that the widths differ by a factor of 4. It is reasonable to suppose that a pair of two such thick or thin lines represents a single digit. The number underneath the barcode is provided so that people can read it and confirm the barcode. Machines themselves do not read this number.

By reading-in the maker code and product code, computers can look up and display a corresponding price from a database. There are also some barcodes which include the price directly.

When a barcode is enlarged with a magnifying glass, the result resembles the image shown in Figure 2. For example, the numeral 3 has a striped pattern with 1 white, 4 black, 1 white and 1 black lines. The numeral 4 has a striped pattern with 2 white, 3 black, 1 white and 1 black lines. One numeral occupies 7 units of width (known as modules), which are arranged in stripes in a white, black, white, black order. Denoting their widths by a, b, c and d, and relating these mathematically, the problem is to find integer solutions satisfying $a + b + c + d = 7$ under the condition that $1 \leq a, b, c, d \leq 4$.

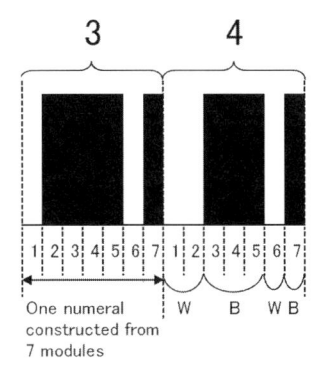

Figure 2: The Structure of the Symbols

2. Enumeration Method

I mentioned that 7 modules are used to represent one numeral, so let's try and think about how many different characters can actually be expressed with 7 modules.

This answer can be obtained effortlessly by using the formula for 'duplicated combinations' which is presented as a 'compiled application' and covered under the topic of combinations and permutations in the chapter on probability and case counting in high-school mathematics A. One cannot, however, guarantee to remember this formula, so after revising it, let's try thinking about the problem from scratch as if we had absolutely no knowledge of it.

What should you do when you have completely forgotten a formula? In such a situation there's nothing for it but to write out all the possible cases that can occur. Figure 3 shows a tree diagram produced by counting them up. This appears simple, but unless the system is written out in an orderly way, oversights will occur. The number of modules, 7, acts as a bound when writing out the cases. One can pretty much give up on trying to write out how many patterns could be produced with 8 or 9 modules of white, black, white, black stripes.

Before we move on and discuss duplicated combinations, let's revise combinations and permutations.

[Question 1] How many ways can 3 people be selected from a group of 10 people and lined up?

There are 10 ways of choosing the first person, then the next is chosen from a group reduced by 1 person, thus containing 9 people, and the

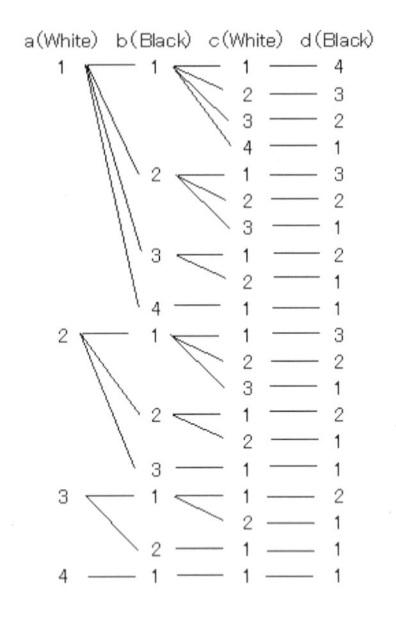

```
             a(White) b(Black) c(White) d(Black)
              1 ┌──── 1 ┌──── 1 ──── 4
              │        │      2 ──── 3
              │        │      3 ──── 2
              │        │      4 ──── 1
              │        │ 2 ┌── 1 ──── 3
              │        │    │  2 ──── 2
              │        │    │  3 ──── 1
              │        │ 3 ┌── 1 ──── 2
              │        │    │  2 ──── 1
              │        │ 4 ─── 1 ──── 1
        2 ┌── │ 1 ┌── 1 ──── 3
          │   │     │  2 ──── 2
          │   │     │  3 ──── 1
          │   │  2 ┌── 1 ──── 2
          │   │     │  2 ──── 1
          │   │  3 ─── 1 ──── 1
        3 ┌── 1 ┌── 1 ──── 2
          │        │  2 ──── 1
          │     2 ─── 1 ──── 1
        4 ──── 1 ──── 1 ──── 1
```

Figure 3: Tree Diagram

next from a group of 8, so there are $10 \times 9 \times 8 = 720$ different ways. If this is expressed using factorials, it may be then be rewritten as a permutation equation.

$$10 \times 9 \times 8 = \frac{10 \times 9 \times \cdots \times 1}{7 \times 6 \times \cdots \times 1} = \frac{10!}{7!} = \frac{10!}{(10-3)!} = {}_{10}P_3$$

In general, the total number of ways of selecting and lining up r things from a collection of n is,

$$_nP_r = \frac{n!}{(n-r)!} \; .$$

[Question 2] How many ways can a combination of 3 people be selected from a group of 10 people?

The result of Question 1 can be used to answer this second question since it reveals the number of permutations by which 3 people can be selected *and lined up* from among 10 people. When the 3 people selected are placed in order, the result is a permutation, but when they are merely selected, the order is not important. The number of ways of ordering 3 people, 3!, is the number of duplicated combinations in each case, so the number of permutations $_{10}P_3$ divided by 3! is equal to the number of combinations, *i.e.*, there are $\frac{10 \times 9 \times 8}{3 \times 2 \times 1} = 120$ different combinations. If this is expressed using factorials, it may then be written as a combinatorial equation, $\frac{_{10}P_3}{3!} = \frac{10!}{(10-3)!3!} = {}_{10}C_3$. In general, the total number of ways of extracting a combination of r things from among n things is $_nC_r = \frac{_nP_r}{r!} = \frac{n!}{(n-r)!r!}$.

3. Duplicated Permutations and Duplicated Combinations

The discussion above covers some basic points learned in high-school mathematics, but combinations and permutations each have extensions known as duplicated combinations and duplicated permutations, respectively, *i.e.*, one may seek the number of combinations or permutations when duplicates are permitted.

While some topics are included in the syllabus as fundamentals and should be learned, others are summarized as extension problems. Since the barcodes we are discussing this time constitute a combinatorial problem with duplicates, allow me to state this issue in detail.

First let's think about duplicated permutations.

[Question 3] How many ways are there to select and line up 3 people from a group of 10 when duplicates are permitted?

If it's difficult to imagine duplicating people, it should be enough to suppose that there are 3 different draws and all the people may participate each time. The number of possible selections each time includes all 10 people, so there are $10 \times 10 \times 10 = 10^3 = 1000$ different ways. In general, there are a total of n^r ways of selecting and lining up r things from among n things.

[Question 4] How many ways can a combination of 3 people be selected from a group of 10 when duplicates are permitted?

As an example, suppose that the 3 people are chosen so that the 3rd person is selected twice, and the 8th person is selected once, then a route like that shown in Figure 4 is conceivable. Looking at it like this unifies the number of ways of selecting 3 people from among the 10 while permitting duplicates, with the route between S and G. There are $(10-1)+3 = 12$ routes between S and G, and the problem becomes one of deciding how to make the 3 steps according to the upwards pointing arrows, or 9 steps according to the right-facing arrow. There are $_{12}C_3 = {}_{12}C_9 = \dfrac{12 \times 11 \times 10}{3 \times 2 \times 1} = 220$ different ways.

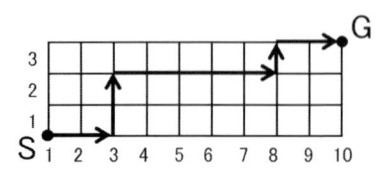

Figure 4: How Many Ways Are There to Make the Route between S and G?

Pay attention to the positions of the numerals written in the bottom line in Figure 4. Since the numerals are on lines extended from a grid, the breadth between 1 and 10 is 9. Understanding the expression $(10-1)+3 = 12$ is particularly important, so let's consider an explanation from another perspective. Question 4 is equivalent to the problem of placing 3 indistinguishable balls into a set of boxes numbered from 1 to 10. Suppose that for the example in Figure 5, there are 2 balls in the '3' box and 1 ball in the '8' box. Removing the box's outer frame reveals that there are 9 empty partitions. With 9 empty partitions and 3 balls, there are a total of 12 locations among which the locations of the 3 balls, or alternatively, the 9 empty partitions, must be chosen. This is a combinatorial problem solved by, $_{12}C_3 = {}_{12}C_9 = 220$. .

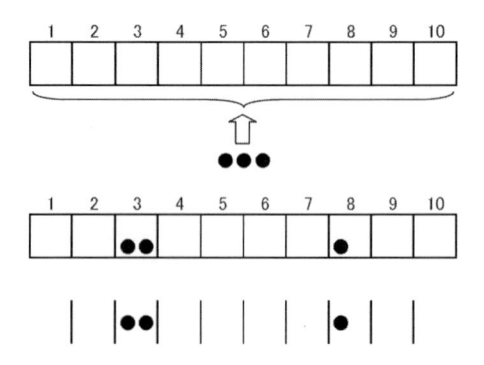

Figure 5: Duplicated Combinations (Alternative Solution)

In general, when performing r selections from a group of n people, when duplications are permitted, the following number of ways are possible.

$$_nH_r = {_{n+r-1}C_r} = \frac{(n+r-1)!}{(n-1)!r!}$$

The number of combinations including duplicates is sometimes written $_nH_r$.

All the necessary preparation is now complete. Let's now think about the combinations involved in barcodes again. Each numeral in a barcode is formed from 7 modules. One numeral is represented by a white + black + white + black pattern of lines. Denoting the widths of these four lines as a, b, c and d modules, the problem is to find the integer solutions of

$$a + b + c + d = 7.$$

However, the width of each black and white line must be at least 1 module, so the maximum module width is constrained. The maximum module width is 4, which may expressed mathematically by seeking solutions constrained by

$$1 \leq a, b, c, d \leq 4.$$

Subtracting the minimum module width from each of a, b, c and d reveals that in the end the problem is to allocate the remaining $7-(1+1+1+1) = 7-4 = 3$ modules. This is equivalent to the problem of finding the total number of ways of selecting a combination of 3 elements from the 4 alternatives a, b, c and d when duplications are permitted. From the formula for combinations with duplicates, the answer is

$$_4H_3 = {_{4+3-1}C_3} = {_6C_3} = \frac{6!}{(6-3)!3!} = 20.$$

There is nothing wrong with looking at the construction of the symbol in Figure 2 and immediately applying the formula for duplicated combinations as shown above. However, many people might not remember the equation without making an error. Therefore, in order to find the solution without making a mistake, wouldn't it be good to count up the different possibilities according to a tree diagram, such as that shown in Figure 3, so as to obtain the solution reliably, without using the formula for duplicated combinations?

4. Forty Different Patterns Are Possible

I explained that using 7 modules, characters can be expressed in 20 different ways, but the inversion of the white, black, white and black pattern (which is a black, white, black and white pattern) can also be used to represent characters in 20 different ways. In total, there are 40 different possible ways of representing characters. Since there are only 10 numerals between 0 and 9, this 4-fold excess may be thought unnecessary, but the leeway is used to prevent mistakes when barcodes are read.

The case when the total number of black modules is an odd number is known as 'odd parity' and the even case as 'even parity.' There are two types of pattern for the maker code (on the left side), *i.e.*, even or odd parity. Only even parity is used for the product code (on the right side). For example, the same number, 3, may be represented with white (1), black (4), white (1) and black (1), in which case there are a total of 5 black modules and the parity is odd, or it may be represented as white (1), black (1), white (4) and black lines (1), in which case there are a total of 2 black modules and the parity is even. Thirty different patterns from JAN (the Japanese Article Number code) are shown in Figure 6.

The maker code (left side) has a white, black, white, black pattern, and the product code has a black, white, black, white pattern. This is related to the reading of the barcode by a scanner, and allows the barcode to be read from either side.

When the number of modules used to represent 1 character is 7, there are 20 possible combinations. If there are 6 modules, then there are 10 combinations. This can be written as a tree diagram, and it is not particularly time consuming to do so. However, when solving the problem of the number of ways characters can be represented when there are 8 modules, it is necessary to rely on the formula after all. The formula is not intended to make examination candidates suffer; it is intended to save the considerable amount of time that would otherwise be spent writing out the cases.

If you are liable to forget the formula, or can only remember it vaguely, it is enough to remember the process behind the introduction of the formulae for permutations, combinations, duplicated permutations and duplicated combinations, as demonstrated through Questions 1 to 4.

In the case of 8 modules the integer solutions of

$$a + b + c + d = 8$$

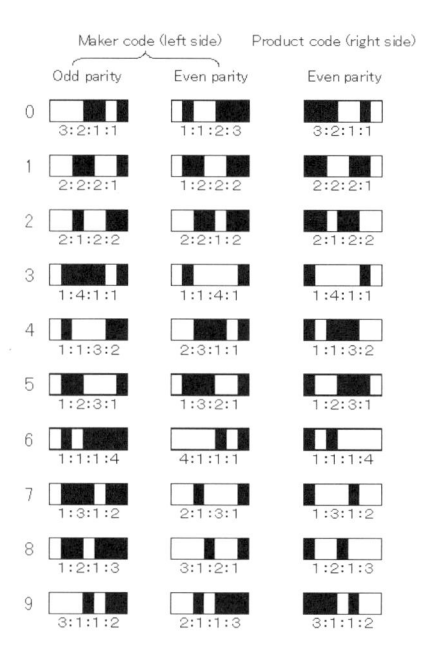

Figure 6: Table Showing the Correspondence between Numerals and Patterns

should be found according to the constraints

$$1 \leq a, b, c, d \leq 5.$$

Since each of a, b, c and d must have at least 1 module each, the result is the number of duplicate combinations of the $8 - 1 \times 4 = 4$ modules among a, b, c and d, *i.e.*, the number of ways is

$$_4H_4 = {}_{4+4-1}C_4 = {}_7C_4 = \frac{7!}{(7-4)!4!} = 35.$$

Considering the black, white, black, white pattern in addition to the white, black, white, black pattern yields 70 alternatives.

The 13th digit of a barcode is a check digit. This is used to confirm that the 12 digits composed of the country code (2 digits), the maker code (5 digits) and the product code (5 digits) are read-in correctly, so it does not represent a numerical value.

The check utilizes a modulus of 10, and I would like to explain how it is calculated. From the 12 digits, the sum of all those in an even position is obtained and then multiplied by 3. The sum of all the digits in odd numbered positions is then added. The result is then subtracted from the smallest multiple of 10 which exceeds its value. The result is the check digit. Let's look at a concrete example, and calculate the specific value corresponding to Figure 1. The 13 digits of this number are 4901306042823, and the meaningful digits are 490130604282. The sum of the digits in even positions is $9+1+0+0+2+2 = 14$. Multiplying this number by 3 yields $14 \times 3 = 42$. The sum of the digits in odd positions is $4+0+3+6+4+8 = 25$, and adding these together yields $42+25 = 67$. The smallest multiple of 10 larger than 67 is 70, and subtracting 67 from 70 yields $70-67 = 3$. The check digit is therefore 3. Owing to this check, the ratio of correctly read-in barcodes is kept high.

References

[1] Y. Nishiyama, Bakodo Shinboru [Barcode Symbols], in *Saiensu no Kaori* [*The Scent of Science*], Tokyo: Nihon Hyoronsha, (1991), 1-9.

[2] Japanese Standards Association, Barcode Symbols for Common Use as Product Codes in Japan, (1985).

CHAPTER 8
Building Blocks and Harmonic Series

Abstract: This chapter presents an explanation of the divergence and convergence of infinite series through the building block problem. The chapter simultaneously expresses the fact that mathematics is not just about manipulating complicated numerical formulae but is also a field in which logical ways of thought are acquired. It is emphasized that in order to overcome university students' aversions to mathematics, lecturers must pour their energies into developing study materials taken from topics relevant to students.

AMS Subject Classification: 00A09, 40A02, 97A20
Key Words: Harmonic series, Center of gravity, Convergence and Divergence, Logarithmic functions, Mathematics education

1. Is it Possible to Stagger Building Blocks by More Than the Width of One Block?

There has been a lot of publicity about how young people avoid and are "allergic" to mathematics. The goal of mathematics is not difficult numerical formulas but a mathematical way of looking at and thinking about things, and I would like to present one example of this. Let us think about the building blocks problem in Figure 1. There are a few building blocks stacked up, and the problem is whether or not it is possible to stack them in such a way that the positions of the bottom block and the top block are horizontally separated by more than the width of one block.

Most people asked this question would immediately answer that it is not possible. I wonder if this tendency to come to a conclusion before even attempting to think about whether something is possible or not is a reflection of the digital age. Sometimes it is possible, sometimes it is not possible, additionally sometimes we do not know. But they hate

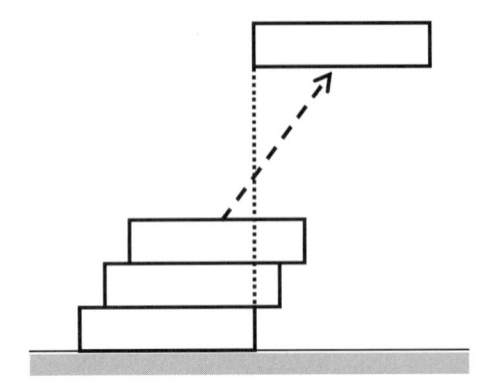

Figure 1: Is it Possible to Stagger Building Blocks by More Than the Width of One Block?

vague answers very much. This is not magic or a trick, I promise that a solution certainly exists. If a person is told to stack building blocks in a staggered way, he or she will stagger them uniformly. But if they are staggered uniformly they will fall down every time. I wonder if this tendency to stack the blocks uniformly is also a manifestation of digital thinking.

We will not obtain a solution immediately. Let us start by looking at the case of two blocks. It is intuitively obvious that the distance they can be staggered is 1/2 of the width of the blocks. So the problem is the third block. Let us hold the third block in our right hand and think about this problem. Most people would try to stack this block on top of the other two but then they always fall down. If your approach does not work it is important to abandon it, and you should search for an alternative approach. To be able to do this, it is necessary to change your way of thinking about the problem.

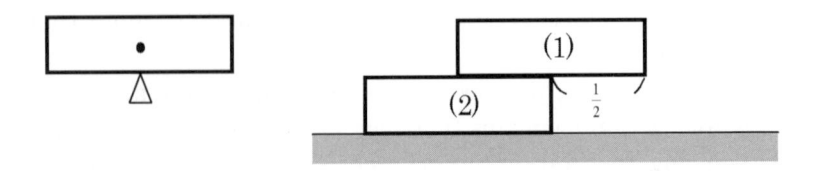

Figure 2: 1/2 Stagger

2. Calculating the Center of Gravity

This is the problem of calculating the center of gravity. Rather than thinking about this problem with a pen and paper, it is surprisingly fast to use building blocks and look for the answer through trial and error.

Here I will give you a hint. Are building blocks best stacked on top of each other? You will probably be perplexed by this hint. That is because of the fixed preconception that it is because we stack them on top of each other that they are building blocks. But building blocks should not be stacked on top of each other; they should be slid under each other. If the third building block is placed at the bottom, and we gradually stagger the first two building blocks on top of the third building block while maintaining the relationship between the first two building blocks as it was, we find that we can stagger the top two blocks by 1/4. In the same way, the fourth block can be placed under the other three and staggered by 1/6, and the fifth block can be placed under the other four and staggered by 1/8. If we add 1/2, 1/4, 1/6 and 1/8 the sum is greater than 1. In other words, we have stacked the blocks in such a way that the position of the top block is horizontally separated from that of the bottom block by more than the width of one block.

While referring to Figures 2, 3 and 4, let us confirm the above approach as a center of gravity calculation using numerical formulae. First of all let us think about building block (1) and building block (2). It is clear that we can only stagger them by 1/2 of the width of a block (Figure 2).

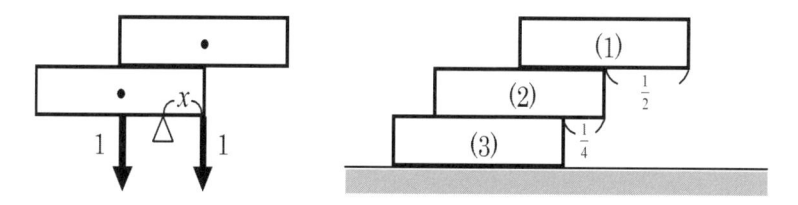

Figure 3: 1/4 Stagger

Next we are going to put building block (3) under the first two blocks, so let us think about the center of gravity of building blocks (1) and (2) together (Figure 3, left). Because building block (3) can be staggered up to the center of gravity, I will obtain the moment, called the stagger distance. Moment is the product of weight and arm length, so the moment of building block (1) (rotated clockwise) is $1 \times x$, the moment

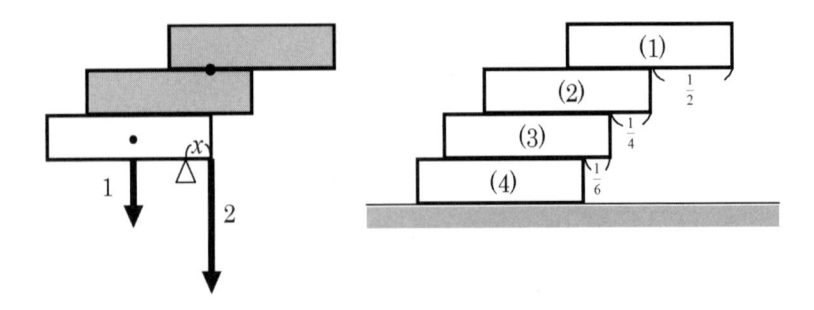

Figure 4: 1/6 Stagger

of building block (2) (rotated anti-clockwise) is $1 \times (\frac{1}{2} - x)$ and because these two values are equal,

$$1 \times x = 1 \times (\frac{1}{2} - x).$$

Solving this equation, we show that $x = 1/4$. In other words, the stagger distance for building block (3) is 1/4 (Figure 3, right).

Next we are going to place building block (4), so let us think about the center of gravity of building blocks (1), (2) and (3) together. Let us obtain this center of gravity from the combination of the center of gravity of building blocks (1) and (2) together and the center of gravity of building block (3) (Figure 4, left). Taking building blocks (1) and (2) together gives a weight of 2. The moment of building blocks (1) and (2) (rotated clockwise) is $2 \times x$, and the moment of building block (3) (rotated anti-clockwise) is $1 \times (\frac{1}{2} - x)$ and because these two values are equal,

$$2 \times x = 1 \times (\frac{1}{2} - x).$$

Solving this equation, we show that $x = 1/6$. In other words, the stagger distance for building block (4) is 1/6 (Figure 4, right).

Let us obtain the general result for the center of gravity of n building blocks. As this is determined by the center of gravity of $(n-1)$ building blocks plus the center of gravity of one building block, $(n-1) \times x = 1 \times (\frac{1}{2} - x)$, therefore $x = \frac{1}{2n}$.

Rearranging this equation we can see that if the stagger position is as follows

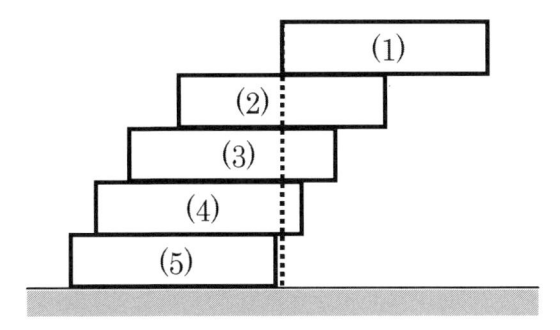

Figure 5: 1-Block Stagger

$$\frac{1}{2}, \frac{1}{4}, \cdots, \frac{1}{2n}, \cdots$$

then the building blocks can be stacked so that they will not fall down. When the progression produced by reciprocal numbers is an arithmetic progression, it is called a harmonic progression. For example, $1, 1/2, 1/3, \cdots$ and $1, 1/3, 1/5, \cdots$ are harmonic progressions. Harmonic progressions are said to have been used in the study of harmonies theory by the Pythagorean School in ancient Greece and the name of harmonic progressions is derived from it. Harmonic series are the totals of harmonic progressions, so we can also write

$$\frac{1}{2} + \frac{1}{4} + \frac{1}{6} + \cdots + \frac{1}{2n} = \frac{1}{2}(1 + \frac{1}{2} + \frac{1}{3} + \cdots + \frac{1}{n}).$$

So now let us calculate the value of this series.

$$\frac{1}{2} = 0.5, \quad \frac{1}{4} = 0.25, \quad \frac{1}{6} \approx 0.167, \quad \frac{1}{8} = 0.125$$

Therefore,

$$\frac{1}{2} + \frac{1}{4} + \frac{1}{6} + \frac{1}{8} \approx 1.042 > 1.$$

So we now know that if we have 5 building blocks we can stagger them by more than the width of one block (Figure 5).

3. Convergence and Divergence

In high school and university differential and integral calculus textbooks, there are chapters on progressions and series. In those chapters the following exercise invariably appears:

CHAPTER 8

$$1 + \frac{1}{2} + \frac{1}{3} + \cdots + \frac{1}{n} + \cdots$$

is divergent, and

$$1 + \frac{1}{2^2} + \frac{1}{3^2} + \cdots + \frac{1}{n^2} + \cdots$$

is convergent.

When n goes to infinity, there are interesting exercises in which sometimes even if the general term of the progression converges to 0 the infinite series diverges. Convergence and divergence can be approximately known by performing integration as follows.

$$\sum \frac{1}{n} \approx \int \frac{dx}{x} = [\log x]$$

Therefore,

$$\sum \frac{1}{n^2} \approx \int \frac{dx}{x^2} = [-\frac{1}{x}].$$

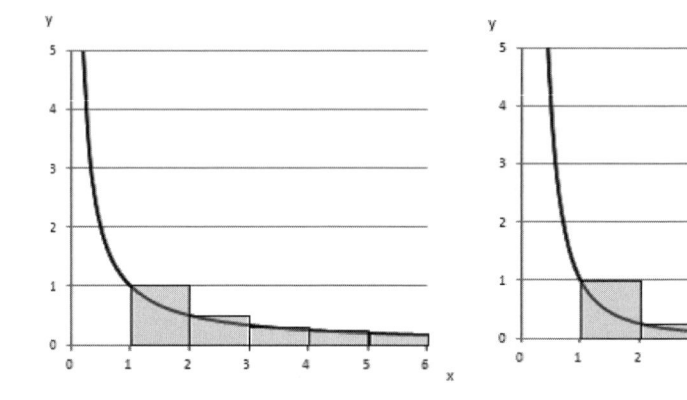

Figure 6: $y = \frac{1}{x}$

Figure 7: $y = \frac{1}{x^2}$

The first of these two equations is in log order and diverges (Figure 6), and the second of these two equations converges (Figure 7). Generally, infinite series of the form $\sum \frac{1}{n^p}$ ($p > 0$) diverge if $p \leq 1$ and converge if $p > 1$. Furthermore, it is known that $\sum \frac{1}{n^2}$ converges to

$\frac{\pi^2}{6}$. Furthermore, whether or not $\sum \frac{1}{n}$ converges is determined by the Cauchy convergence criteria for the progression.

The sum of the first n terms of the progression $a_1, a_2, \cdots, a_n, \cdots$ is defined as

$$S_n = a_1 + a_2 + \cdots + a_n.$$

As for the necessary and sufficient condition for the series $\sum a_n$ to be convergent, if we make N sufficiently large compared to any given positive number ϵ, for all n and m where $m > n > N$ it can be shown that

$$|S_m - S_n| = |a_{n+1} + a_{n+2} + \cdots + a_m| < \epsilon.$$

Assuming that

$$S_n = 1 + \frac{1}{2} + \frac{1}{3} + \cdots + \frac{1}{n},$$

then no matter how big we make n,

$$|S_{2n} - S_n| = \frac{1}{n+1} + \frac{1}{n+2} + \cdots + \frac{1}{n+m}$$
$$> \underbrace{\frac{1}{2n} + \frac{1}{2n} + \cdots + \frac{1}{2n}}_{n \ terms}$$
$$= \frac{1}{2}.$$

So the *Cauchy* convergence criteria are not met. Therefore $\sum \frac{1}{n}$ is divergent. Let us look at this more closely. If we take the number of terms $2n$ as powers of 2 like this: $2, 4, 8, \cdots$, then

$|S_2 - S_1| = \frac{1}{2}$

$|S_4 - S_2| = \frac{1}{3} + \frac{1}{4} > \frac{2}{4} = \frac{1}{2}$

$|S_8 - S_4| = \frac{1}{5} + \frac{1}{6} + \frac{1}{7} + \frac{1}{8} > \frac{4}{8} = \frac{1}{2}$

$|S_{2n} - S_1| = |S_{2n} - S_n| + \cdots + |S_8 - S_4| + |S_4 - S_2| + |S_2 - S_1|$
$$> \frac{1}{2} + \cdots + \frac{1}{2} + \frac{1}{2} + \frac{1}{2}.$$

So we can see that the series diverges.

4. Divergence in Log n Order

I have explained that the harmonic series $\sum \frac{1}{n}$ diverges to infinity, but let us look closely at how quickly $\sum \frac{1}{2n}$ diverges. I used a personal computer to calculate the value of $\sum \frac{1}{2n}$, the total stagger distance. The results were as follows.

When $n = 4$ $\sum \frac{1}{2n} = 1.0417 > 1$

When $n = 31$ $\sum \frac{1}{2n} = 2.0136 > 2$

When $n = 227$ $\sum \frac{1}{2n} = 3.0022 > 3$

So the series does diverge to infinity but at an extremely slow speed. If we now compare $\sum \frac{1}{n}$ with the integration of the function $y = \frac{1}{x}$, we can establish an inequality as follows.

$$\frac{1}{2} + \frac{1}{3} + \cdots + \frac{1}{n} < \int_1^n \frac{dx}{x} < 1 + \frac{1}{2} + \cdots + \frac{1}{n-1}$$

From the fact that $\int_1^n \frac{dx}{x} = [\log x]_1^n = \log n$, we can show that

$$\frac{1}{2}(\log n + \frac{1}{n}) < \sum \frac{1}{2n} < \frac{1}{2}(\log n + 1).$$

So we know that when $n \to \infty$, $\sum \frac{1}{2n}$ diverges in $\frac{1}{2} \log n$ order.

Because one more extra building block is necessary at the bottom, the number of building blocks necessary is actually $n + 1$. Only five building blocks (4+1) are sufficient to stagger the pile of building blocks by the width of one building block, but 32 building blocks (31+1) are necessary to stagger the pile by the width of two building blocks and 228 building blocks (227+1) are necessary to stagger the pile by the width of three building blocks. Figure 8 shows a stack of 32 building blocks, but in practice it is impossible to stack up 32 blocks accurately in this way. This is just a theoretical discussion. Figure 9 is a graph showing the function $y = \frac{1}{x}$ and $y = \log x$, the function resulting from the integration of $y = \frac{1}{x}$. If we rotate the log function 90 degrees clockwise and reverse it horizontally it becomes the building block stacking problem in Figure 8. I will leave it to you to confirm this.

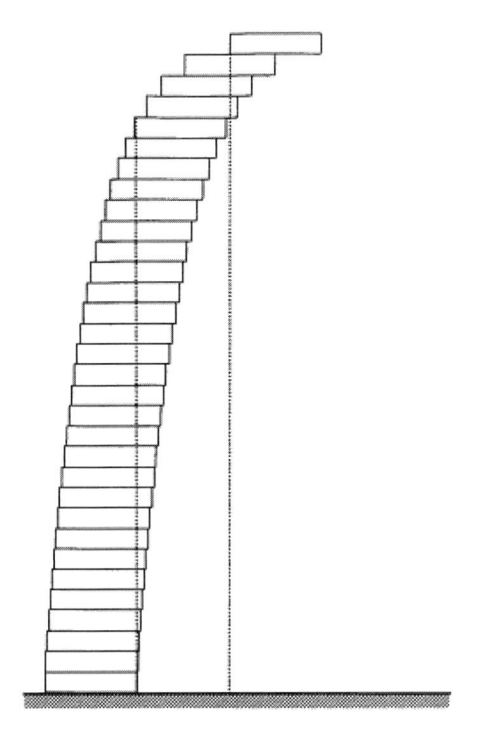

Figure 8: 2-Block Stagger

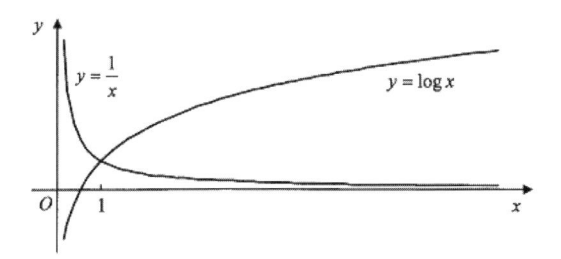

Figure 9: Log Function

That completes the proof. I have shown that the harmonic series $\sum \frac{1}{n}$ describes the solution to the building blocks problem. If we solve these kinds of problems, mathematics should be more enjoyable I think. Mathematics in high school and university progressively becomes more distant from reality and sometimes students come close to losing sight of it. At times like that, students must not forget to apply the problems to reality. The building blocks problem is the problem of the calculation of the center of gravity; it also involves harmonic series and is extremely mathematical. If we limit ourselves to just solving the problem, we do not need to use complicated numerical formulae. The important things are to employ logical ways of thought and to have the ability to change your way of thinking. Ironically, university students doing science subjects cannot solve this building blocks problem. They can prove with numerical formulas that harmonic series diverge to infinity, but they cannot solve the real world problem of the building blocks. This is a blind spot in modern education.

I leaned about the building blocks problem from a 1958 work by George Gamow [1]. He was both a researcher and educator and it appears that he was of the opinion that students will not get excited about mathematics if the teacher is not excited about it. If you would like to confirm the solution to the building blocks problem but you do not have any building blocks at hand, you could try doing it with ten volumes of an encyclopedia or ten video tapes.

Reference

[1] G. Gamow, M. Stern, *Puzzle-Math*, The Viking Press Inc., USA, (1958).

CHAPTER 9
Playing with Möbius Strips

Abstract: Möbius strips are well known in topology as curved surfaces with no inside or outside. The degree of twist in the loop is particularly important, that is, whether it is an odd or even multiple of 180 degrees. This article shown how interesting mathematics can be, by splitting Möbius strips into halves or thirds using a pair of scissors, and contains a hint for unravelling the structure of space.

AMS Subject Classification: 00A09, 51H02, 97A20
Key Words: Möbius strip, Topology, Non-Euclidean geometry, Endless tapes

1. A Single Sheet of B4 Paper

This time, let's talk about Möbius strips. Möbius strips, also known as Möbius loops, are discussed in topology but they also include very evocative aspects for the general public, so I always make a point of taking them up during my seminar time. I'd be pleased if those readers who are not familiar with Möbius strips would try them out.

To begin with, prepare a single sheet of B4 copy paper. From experience, a sheet with height 257 mm and width 364 mm is best. Divide this into 6 equal parts as shown in Figure 1, making 6 long thin paper strips. These 6 strips will be used for experiments.

So let's try out some experiments with Möbius strips. First, make one Möbius strip. For those who don't know about Möbius strips, they can be made by gluing, just like a usual loop, except that Möbius strips have a 180 degree twist (Figure 2). There are two ways to twist by 180 degrees, clockwise and anticlockwise, but either direction is fine. Just by twisting the strip through 180 degrees and gluing it together, a weird world begins to open up.

Having checked that the Möbius strip is made correctly, let's think about what will happen if the strip is split in half. These days students

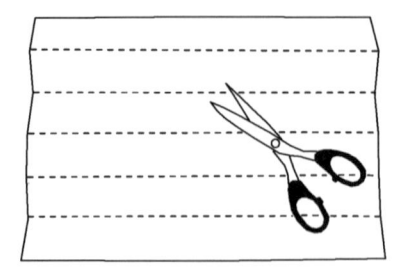

Figure 1: B4 Sheet Divided into 6 Equal Parts

(a) Normal loop

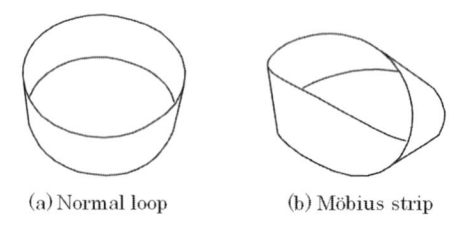

(b) Möbius strip

Figure 2: Normal Loops and Möbius Strips

seem to prefer cutting the loop with a pair of scissors before even thinking about the result, so those of you with scissors should be warned that you will lose points if you don't first try to imagine the result. Many people predict that cutting the loop will separate it into two pieces. Next comes the time to cut the loop and see. As an unexpected surprise, when the Möbius strip is cut, rather than separating into two pieces, one large loop is produced (Figure 3). People who didn't know about Möbius strip are surprised, and the experience is also useful for opening up the eyes of young people who have a dislike for mathematics.

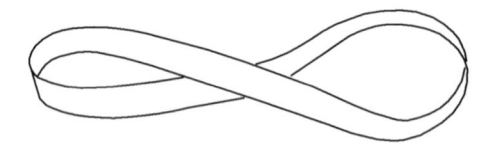

Figure 3: When a Möbius Strip Is Split in Half

The prediction that the strip would separate into two is way off. Let's think about why it forms a one big loop rather than separating into two. Since we know the difference between a normal loop and a

Möbius strip with a 180 degree twist, make a normal loop and a Möbius strip, and try running a line along each using a pencil (Figure 4). With a normal loop, when a line is run along the outer surface it doesn't reach the inner surface, but with a Möbius strip the line runs along the inside and the outside. For those of you discovering the Möbius strip for the first time, the expectation that it would be the same as a normal loop makes this alone an impressive phenomenon.

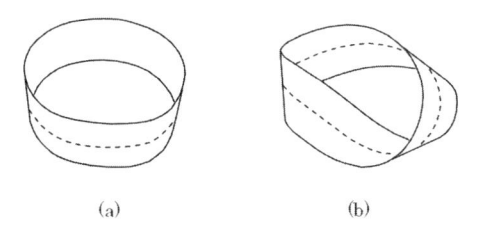

Figure 4: Running a Line in Pencil

2. Odd and Even Multiples of 180 Degrees

The work so far was round 1. Some people know that cutting a Möbius strip produces one long thin loop, but this story has a continuation. What happens when the loop produced so far (Figure 3) is again cut into two?

Let's make a prediction. At this point the prediction that one even longer loop will be produced is common. Is it perhaps a characteristic of the current younger generation that they are easily influenced by previous results. One person says that a large loop will be produced and then everyone responds that they think the same thing. Let's think about it carefully. Here's a hint. The thing we're trying to cut is a Möbius strip. Is it that the desire to see one big loop is stronger, or is it that there are few alternative predictions? The result is that two loops are produced as shown in Figure 5, and they are linked. People's predictions are again mistaken.

Reviewing the results above yields the following. The degree of twist in the loop is particularly important. A normal loop without any twist has an inside and an outside, but the Möbius strip with its 180 degree twist is a 'curved surface with no inside or outside' since the inner and outer surfaces cannot be distinguished. Furthermore, it can be written on the board that a loop with an even number of 180 degree twists (0 degrees, 360 degrees, 720 degrees...) has an inside and an outside, while

Figure 5: When the Loop Is Cut Again

a loop with an odd number (180 degrees, 540 degrees, 900 degrees...) has no inside or outside.

If this were known, it should at least have been possible to predict that the loop in Figure 3 would separate into two pieces because it is twisted by 360 degrees and has an inside and an outside.

3. Splitting a Möbius Strip into Thirds

Let's make another Möbius strip, and this time let's think about what will happen if it is split into thirds. Running a pencil line around the strip reveals information significant for making a prediction, so let's do this. An approximate division into thirds is sufficient. When the pencil line running around the Möbius strip reaches the other side, just like the halfway line, there is no true reverse side. The line is offset, but ignoring this and continuing the line all the way round, it eventually reaches its initial point. Change the color of pencil and draw one more line. In this way the state divided into thirds can be confirmed. If it still seems to be difficult to make a prediction, applying color in the way shown in Figure 7, and thus dividing into different colored regions should help. It appears that the center strip (white) and the two outer strips (black and gray) are different pieces, although the black and gray parts are connected.

Splitting the Möbius strip into thirds yields a small loop and a large loop, which are linked (Figure 8). The small loop is a Möbius strip. The large loop is a surface with an inside and outside, and has a 360 degree twist in it.

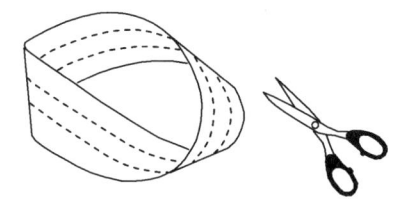

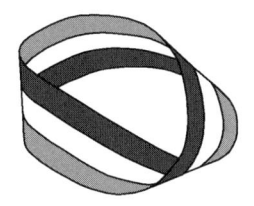

Figure 6: Möbius Strip Split into Thirds

Figure 7: Applying Color

Figure 8: The Large Loop and Short Loop are Linked.

4. A Hint for Unravelling the Structure of Space

Who devised the Möbius strip, and for what purpose? A.F. Möbius (1790-1868) was a German mathematician and astronomer active in the 19th century. He proposed it, so the Möbius strip is named after him. He was also an astronomer so he had an interest in the structure of space, and it is assumed that he devised the strip while exploring various notions about the boundaries of the universe.

Stepping back for a moment into an earlier age, it was thought that the world was flat. People thought that if you went to the edge where no one had been, there was a cliff that you would fall off. The fact that the world is round was demonstrated by Columbus, who discovered the 'New World' in 1492. Afterwards, Magellan (1480-1521) made the first circumnavigation of the world, thus proving the fact.

This also had a significant impact on geometry. Up until that time geometry was essentially Euclidean, and it was thought that all problems might be solvable with this well-integrated model. There was an approximate contemporary of Möbius known as Lobachevsky (1792-1856). He was a Russian mathematician who, while pursuing research into parallel lines, validated a formulation of non-Euclidean geometry and presented his results.

Non-Euclidean geometry is born from a rejection of the axiom of parallel lines. In this geometry, the axiom of parallel lines is replaced by the notions that 'for a given straight line on a plane there are at least two straight lines no different from the given line that passes through a point which is not on the given line,' 'the sum of the internal angles of a triangle is at least the sum of two right angles' and furthermore, 'straight lines are finitely closed, and divergences from Euclidean geometry can also be seen in the ordering of points on a straight line,' etc.

For example, think about a triangle NAB formed by the North Pole and two points on the equator. In this case the sum of the internal angles is larger than the sum of two right angles. The sum of the internal angles of a triangle drawn on a notepad is the sum of two right angles, but triangles on the Earth's surface do not accord with Euclidean geometry. There is a relationship in the sense that non-Euclidean geometry is locally Euclidean.

When Columbus and Magellan proved that the Earth is round, the mistake regarding the edge of the world was resolved. The ancient mystery regarding the nature of the boundary of the universe on the other hand remains. Space is vast, and being unable to reach the boundary, humans cannot resolve this problem absolutely. There is a theory that space, like the Earth, is closed.

Until recently, it was thought that light traveled in straight lines, but the person who rejected this in an attempt to clarify the nature of space was Einstein (1879-1955). He established a new 4-dimensional model of the universe according to the theory of special relativity based on the principles of relativity and the constancy of the speed of light in an inertial reference frame, discarding the notions of absolute space, absolute time, the ether and so on occurring in Newtonian dynamics. The unity of mass and energy are derived from this theory. Beginning in 1907, it was attempted to expand the theory of relativity to gravitational fields, leading to completion of the general relativity theory in 1915.

Light bends according to gravity. The light released into space from the Earth experiences the effects of gravity and will at some point return. The theory that space does not have a boundary but is closed was therefore composed.

Regarding space with 3-dimensional coordinates, adding a further axis for time yields a 4-dimensional coordinate frame. On the Earth 3-dimensional coordinates are sufficient, but at the astronomical level 4-dimensional coordinates become necessary. Euclidean geometry and Newtonian dynamics are no good for understanding the boundaries of space, and non-Euclidean geometry and Einstein's theory of relativity become necessary.

5. Can We Go to a 4-Dimensional World?

It's somewhat metaphorical but let's try thinking about dimensions. Caterpillars and ants can only move in straight lines. They are 1-dimensional animals that only know a single route. Whirligig beetles, which float on water and can only glide on the water's surface are restricted to planar movement and are 2-dimensional animals. People are 3-dimensional animals capable of conceiving straight lines, planes and spaces. The only 4-dimensional animals capable of perceiving the speed of light as finite and grasping the structure of space would be extraterrestrial.

We can understand that the animals living at each dimensionality are at a lower dimensional level than ourselves, but we cannot comprehend a higher level of dimensionality. For example, we are 3-dimensional animals, so while we can comprehend ants which are 1-dimensional animals and whirligig beetles which are 2-dimensional animals, we cannot imagine 4-dimensional space. This is the same relationship according to which 1-dimensional ants cannot comprehend planes and whirligig beetles cannot comprehend spaces.

Among the many ideas about how to express a 4-dimensional coordinate system in a 3-dimensional world, the Möbius strip is often cited. Even when living in a 2-dimensional world, mounted on a Möbius strip it is possible to reach the other side of the surface. If the relationship between the inside and the outside is taken as a concept going beyond the notion of a plane, then the Möbius strip is a bridge from 2 dimensions to a quasi-3-dimensional world. This embodies the expectation that if a device like this could be manufactured, it might perhaps provide an interface from 3 dimensions to a 4-dimensional world.

6. Application to Endless Tapes

Möbius strips have been proposed as one method for clarifying the structure of space, but besides academia, their wonderful geometrical properties have also been applied in our daily lives, as endless tapes. Musical recording media are moving from CDs (compact discs) to hard disk devices such as iPods (Apple), but at the peak of the analogue generation magnetic tapes were in use.

Magnetic tape kept inside a plastic case is known as a cassette tape. A cassette tape has two sides, A and B. When each side had finished playing the tape had to be turned over or rewound. A way to save the time spent idling under these operations is to connect the tape with a twist of 180 degrees like a Möbius strip. These endless tapes were used in places like storefronts where it is necessary to play them repeatedly. Möbius strips were also applied to computer printer ribbons, although these devices are now to be found in museums.

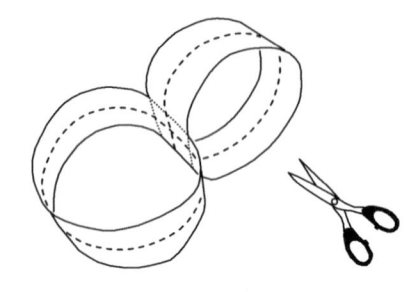

Figure 9: Cross-shaped Connection

After my students have thoroughly played with the Möbius strip, I always make a point of presenting the following quiz to them. Two normal strips are prepared. They are firmly glued, and attached in a cross shape as shown in Figure 9. Well then, what happens when this is split into two along the dotted line?

This is also interesting to predict. Since the results so far have produced large loops, large loops linked with smaller loops and so on, predictions extending these results are common. After gathering all the students' predictions, I have them cut with a pair of scissors and see. Lo and behold, a regular quadrilateral is produced (Figure 10). The fact that a square is produced from the two loops, thus yielding a 2-dimensional surface from a 3-dimensional solid, is an intriguing curiosity. No one expects the result to be planar.

The square thus produced can also be used to make two normal loops. It is helpful to imagine the images of the cutting process played in reverse. The square in Figure 10 passes through an intermediate form like that shown in Figure 11, and ends up in a state like that shown in Figure 9. Essentially, this can be understood by employing a flexible way of thinking.

When performing this quiz, one of my students asked what happens when 4 loops are connected? I hadn't tried this, so I attached normal

Figure 10: Square

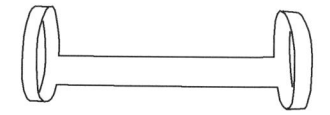

Figure 11: Intermediate Form

loops in 4 crosses as shown in Figure 12 and used a pair of scissors to cut them and see. This time the result was a boxed array (Figure 13). What happens when 8 are connected together and cut? Such interesting topics arose one after the other, when my students played with the Möbius strip for the first time.

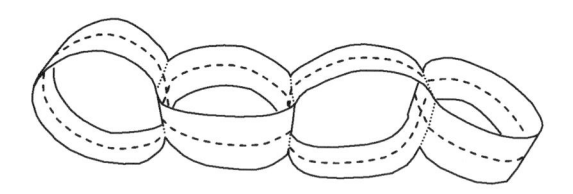

Figure 12: When Normal Loops in 4 Crosses Are Cut

Figure 13: A Boxed Array Is Formed

CHAPTER **10**

A General Solution for Multiple Foldings of Hexaflexagons

Abstract: This article explains hexaflexagons: how to make them, how to operate them, and their mathematical theory. Hexaflexagons are known to be surfaces with no inside or outside, similar to Möbius strips. Referring to the articles of Gardner and Madachy, the author discovered and describes here a general solution for multiple foldings of hexaflexagons.

AMS Subject Classification: 00A08, 00A09, 97A20
Key Words: Hexaflexagon, Möbius strip, Topology, Paper folding

1. Surfaces with No Inside or Outside

In the December 1990 edition of the *Basic Mathematics* magazine, I introduced a handmade puzzle known as a hexaflexagon under the title 'Folding Paper Hexaflexagons' [4]. It has been 10 years since then. The theoretical work related this puzzle has advanced significantly, and the puzzle is now understood. A new folding technique has in fact been developed. I'd like to introduce this puzzle to those readers who do not know it, and explain its close relation to mathematics.

The puzzle was devised in 1939 by the English mathematician Arthur H. Stone, and is known as a hexaflexagon. Perhaps because the name hexaflexagon sounds unfamiliar, it is often called an 'origami hexagon' or 'pleated origami' in Japanese, and all refer to the same thing.

The 'hexa' in hexaflexagon means six, and 'flexagon' indicates something that is flexible, easy to bend, and can take many shapes. There are flexagons in shapes other than hexagons, such as tetraflexagons, which are square, but the most interesting, from both a theoretical and practical perspective, is the hexaflexagon.

I first heard how interesting this puzzle is in 1985, from a report by Shin'ichi Ikeno that appeared in *Mathematical Science* [2]. In fact the puzzle isn't new to Japan, and resembles an old toy known as a *byoubugai*. The puzzle is made of paper and has a hexagonal shape. The hexagon is constructed from six triangles, and by squeezing two adjacent triangles between the thumb and index finger as shown in Figure 1, a new face can be revealed from the center.

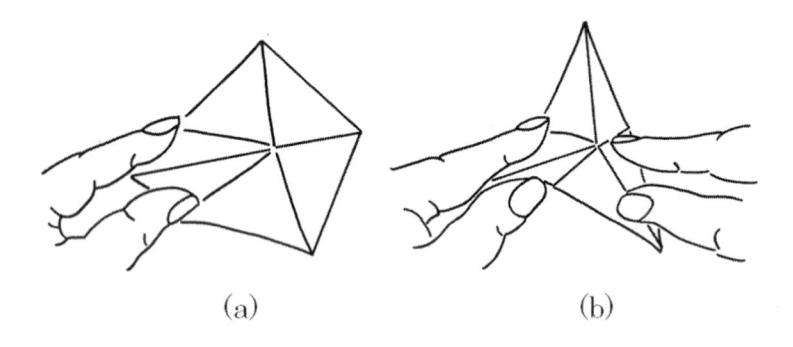

(a)   (b)

Figure 1: Revealing a New Face

Figure 2 illustrates a face which from a topological perspective has no inside or outside. Known as a Möbius strip, it is a normal loop glued together with a 180 degree twist, and was devised by the German astronomer A.F. Möbius (1790-1869). The twist may be to the left or right, and yields a connected surface for which an inside and outside cannot be distinguished. The Möbius strip involves a 180 twist but the hexaflexagon is made with a 540 degree twist; 540 degrees is 3 times 180 degrees. In general, gluing together a strip with an odd multiple of 180 degree twists yields a surface with no inside or outside, while an even multiple yields a face with an inside and an outside.

2. Three Face Folding

Now, fundamentals are important. The 3-face fold used for the hexaflexagon is a basic among basics, so I'd like for the reader to master it completely.

Ten equilateral triangles with sides of 6 cm are lined up sideways as shown in Figure 3(a). It should be possible to draw a diagram of this complexity with a ruler and compasses. The right-hand edge of the 10 triangles is for gluing, so in fact 9 triangles are involved in the

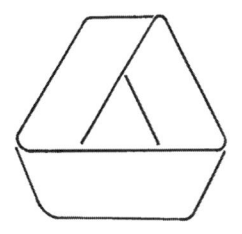

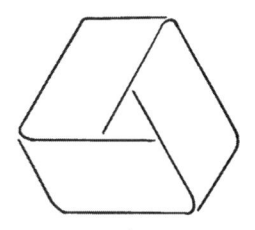

(a) Möbius strip (180 degree twist)   (b) Hexaflexagon (540 degree twist)

Figure 2: Surfaces with No Inside or Outside

puzzle. The triangles each have inner and outer faces, so there are a total of $9 \times 2 = 18$ triangles. The hexaflexagon on the other hand, is composed of 6 triangles; $18 \div 6 = 3$, so mathematically, it is natural that it constitutes a 3-face folding.

While it may tally mathematically however, the appropriate arrangement of the triangles is key, and is explained below. Let's focus on the correct folding technique first. Make a valley fold along line $a-b$ (Figure 3(b)), a valley fold along line $c-d$ (Figure 3(c)), then without restricting the <glue> part, make a valley fold along line $e-f$ and glue (see Figure 3 (d)). This involves 3 valley folds which is a twist of $180 \times 3 = 540$ degrees.

Squeezing two adjacent triangles of the glued hexaflexagon in the way shown in Figure 1 causes a new face to appear naturally from the center. If it doesn't appear, try sliding back one of the triangles (at 60 degrees from the central angle) without pulling too hard. If it still doesn't appear, then the hexaflexagon was constructed incorrectly and should be remade according to Figure 3.

Let's confirm that the hexaflexagon performs correctly. Fill in the numbers on the hexagonal face as shown in Figure 4, with '1' on the first face, '2' on the next face to appear and '3' on the next. The cyclic order $1 \to 2 \to 3 \to 1 \to 2 \to 3$ of faces appearing is characteristic.

It is interesting to know the actual positions of the faces numbered 1 to 3. Figure 5 shows a hexaflexagon that has been peeled open and spread out again. It shows the flaps of paper with the numbers (1), (2) and (3) written on the underside. The same numbers are not written on continuous areas, but pairs of two are lined up in equally spaced positions on both sides. Considering the folding relationship shown in Figure 1, for every fold, the triangle in Figure 5 is offset by 2 steps. The hexaflexagon is thus a single long thin segmented face seen in a

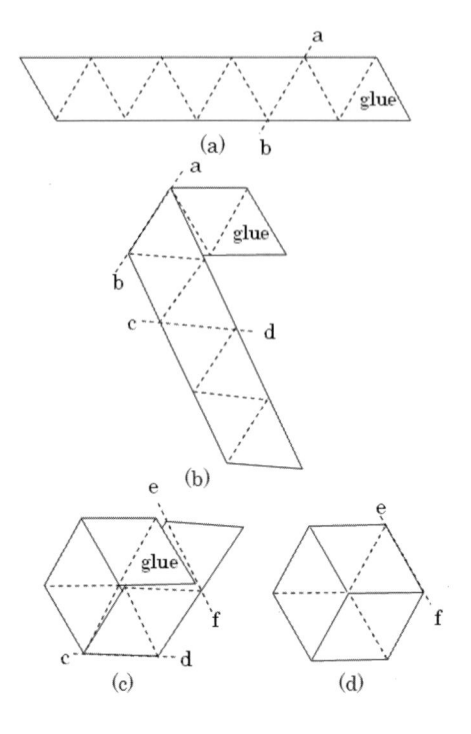

Figure 3: Folding Order (3-Face Folding)

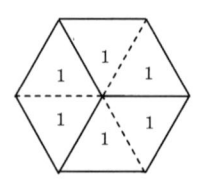

Figure 4: Numbering the Faces

staggered manner.

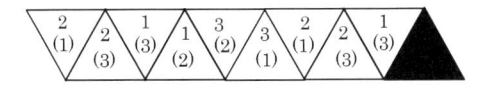

Figure 5: The Relationship among the Three Faces

3. Martin Gardner's Paper Templates

The report from 1990 introduced above only discussed 3-face folding. My own interest moved on to the question of whether there are folding methods for larger numbers of faces.

Martin Gardner's *The Scientific American Book of Mathematical Puzzles and Diversions* contains an article introducing the hexaflexagon, and on page 25 there are paper templates for between 4- and 7-face folds [1]. The book only contains paper templates for folding diagrams and doesn't include an explanation of the folding technique. Since no solution is printed, it is necessary to find one through one's own efforts. After repeatedly failing many times, and thinking to myself 'not like this... not like that...,' I eventually succeeded in making these models.

When making hexaflexagons with many faces ($n \geq 4$), it becomes clear that not only the theory, but also the actual paper used for construction, techniques for making diagrams and so on also become problematic. When I first heard of the puzzle in around 1985, I used drawing paper, a ruler and compasses to make the diagrams. This is reasonable when handling only a 3-face fold, but as the number of faces increases, the accuracy of the diagrams becomes more of a requirement. The lead in a pencil is 0.3 mm, and the graduations on a ruler are in units of 1 mm, so no matter how carefully the diagram is drawn the error in a handdrawn diagram must be at least around 0.1 mm. Even supposing that the error in a single triangle is 0.1 mm, when 10 triangles are included the error accumulates and reaches 1 mm. When making a 12-face fold, the number of triangles is 37, so the error is 3.7 mm and cannot be ignored. Also, drawing paper was used at first, but while drawing paper appears to be strong, it is surprisingly useless, often tearing during bending and folding.

Based on these experiences I abandoned the ruler and compasses, and instead made the diagrams using the language known as Visual Basic. When a computer is used, the hand drawing error of 0.1 mm and accumulated error of 3.7 mm do not arise and the result is considerably more accurate. Drawing paper is weak when it comes to bending and folding, so normal photocopy paper was used instead. I suppose that the quality of the fibrous material must be different. Lastly, although numbers were first written on the faces in order to distinguish them, I gradually realized that classifying them by color was more appealing and therefore filled them in using colored pencils. Copy paper is thin however, and the color shows through to the other side, so colored origami paper was attached using glue.

4. Reduction to a Fundamental Pattern

Now, allow me to explain how I achieved the 4- to 8-face foldings. Figure 6 shows the arrangement of a paper template for a particular representative example. The black triangle is used as an overlap for gluing, and has no relation to the actual appearance.

The 6-face folding is comparatively easy, so let's begin there. The template for the 6-face folding is simply two templates for the 3-face folding (Figure 3(a)) glued together side by side. The number of triangles is 18, but there is one extra used as an overlap (colored black), so the total is in fact 19. If the model is folded from the right-hand edge in an orderly manner using a right twist rule, it is the same as the 3-face folding. The 6-face folding may thus be achieved by applying the 3-face folding.

Long straight paper strips such as the 3-face folding and the 6-face folding are referred to as 'straight models' by Joseph Madachy [3]. These straight models are formed according to the following equation.

$$n = 3 \times 2^p, \quad (p \geq 0, 1, 2, \cdots)$$

Substituting the shown values for p yields $n = 3, 6, 12, 24, \cdots$, meaning that the 3-face, 6-face, 12-face and 24-face foldings are possible with this method. Indeed, $n = \infty$, so a model with an infinite number of faces is also possible in theory.

The basis of the remaining models is a reduction to the fundamental pattern of the straight models (Figure 7). Regarding the folding technique, let's look at the 4-face and 7-face foldings.

For the 4-face folding, by taking the 3 parts below the dotted lines in bottom-up order, and folding using a right twist rule 3 times, the 3-face folding may be applied. The layered parts are indicated in gray, and since these parts form a new face, they are marked '4.' The 4-face

HEXAFLEXAGONS 111

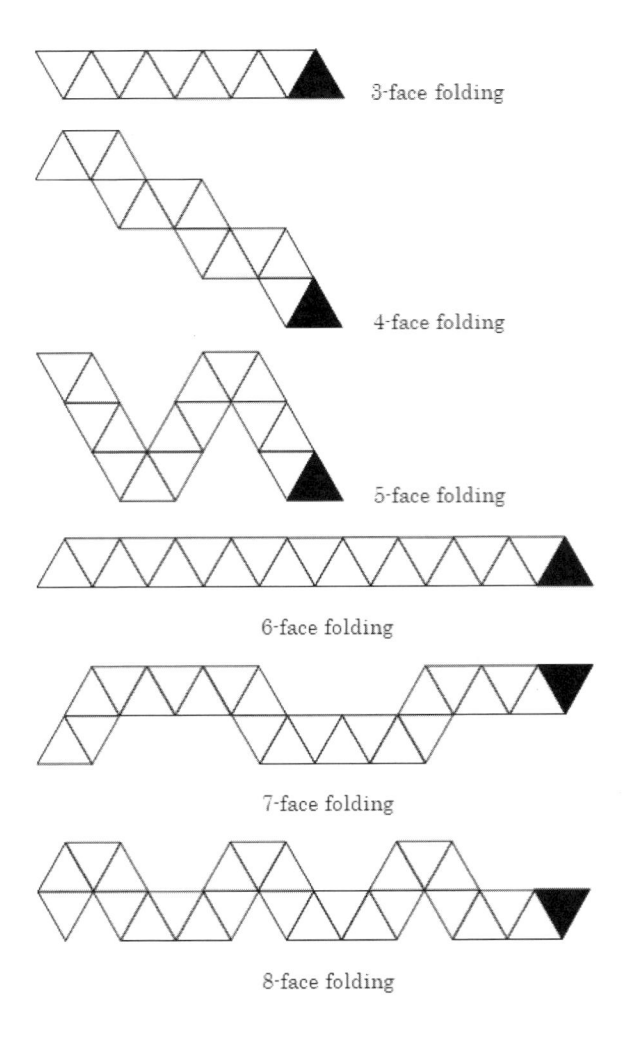

Figure 6: Three- to 8-Face Foldings (Paper Templates)

112 CHAPTER 10

4-face folding → 3-face folding

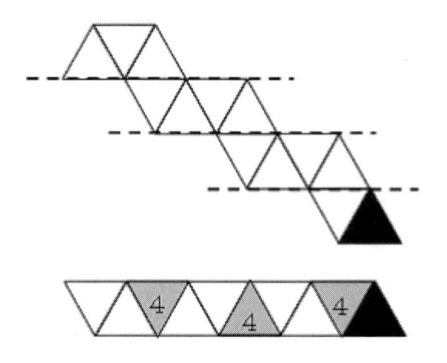

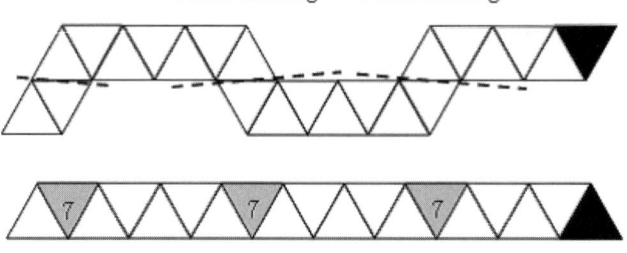

4-face folding layered state

7-face folding → 6-face folding

7-face folding layered state

Figure 7: Reduction to the Fundamental Pattern

folding may be completed by applying the 3-face folding to the layered state.

For the 7-face folding, by taking the 3 parts inside the dotted lines in right-left order, and folding using a right twist rule 3 times, the template for the 6-face folding may be applied. The number '7' was written on the layered parts. This 6-face folding template may be completed by transforming it and applying the template for the 3-face folding. In short, this is a 7-face folding → 6-face folding → 3-face folding procedure.

5. Transition Diagram

If the model is folded up as above, it is certain that only the target number of faces will be revealed. What, however, is the order in which the faces appear? The answer may be found by referring to the transition diagram in Figure 8. I drew up this diagram by referring to the work of Joseph Madachy [3].

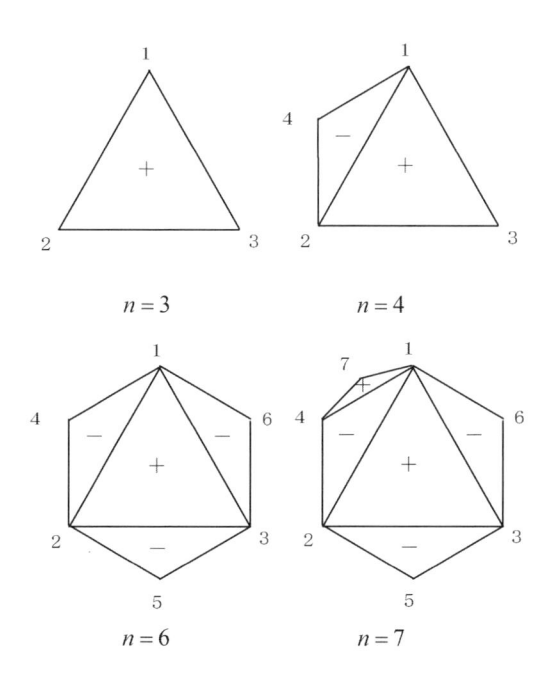

Figure 8: Transition Diagrams

In the case of the 3-face folding ($n = 3$), the transition diagram is expressed as a triangle. The numbers 1, 2 and 3 written at the tips of the triangle are the numbers of the faces. There is a plus (+) symbol inside the triangle, and this signifies that the face numbers cycle in an anticlockwise manner $1 \to 2 \to 3$.

In the case of the 4-face folding ($n = 4$), a new triangle has been added to the transition diagram of the 3-face folding ($n = 3$) in the area between tips 1 and 2. This is the triangle related to the new face with number 4. The triangle is marked inside with a minus ($-$) symbol, signifying that the face numbers cycle in a clockwise manner $1 \to 2 \to 4$. There are thus two cycles existing in the 4-face folding: the plus (+) cycle $1 \to 2 \to 3$, and the minus ($-$) cycle $1 \to 2 \to 4$. For example, to go from 3 to 4 it is not possible to advance directly through $3 \to 1 \to 4$. Instead, by advancing in the $1 \to 2 \to 3$ plus (+) cycle through $3 \to 1 \to 2$, and then advancing in the $1 \to 2 \to 4$ minus ($-$) cycle through $2 \to 4$, the target can be reached. In this case, 2 acts as a relay point.

In the case of the 6-face folding ($n = 6$), three triangles are added to the transition diagram of the 3-face folding. Around the plus (+) cycle $1 \to 2 \to 3$, there are three minus ($-$) cycles $1 \to 2 \to 4$, $2 \to 3 \to 5$, and $1 \to 6 \to 3$.

In the case of the 7-face folding ($n = 7$), a triangle with a plus cycle $1 \to 7 \to 4$ is added to the outer edge of the transition diagram for the 6-face folding ($n = 6$).

The transition diagram for an n face folding thus complies with an n sided polygon, and this n sided polygon is partitioned into $n-2$ triangles such that the adjacent triangles have a different symbol (indicating the cycle direction). By constructing the transition diagram, the operations needed to reveal a particular face may be performed smoothly.

6. General Solution for Multiple Foldings

Paper templates for the 3- to 8-face foldings are shown in Figure 6, and an explanation summarizing the folding processes is shown in Figure 7, but how should foldings for 9 or more faces be handled? Allow me to explain.

To begin with, the existence of the fundamental pattern of the straight model is just as stated above. Expressed as $n = 3 \times 2^p$, $(p \geq 0, 1, 2, \cdots)$ the values are $n = 3, 6, 12, 24, \cdots$ and so on. The templates for the 12- and 24-face foldings are long thin strips. So what happens with larger values of n? Just as the 7- to 11-face foldings may be reduced to the fundamental pattern with the 6-face folding as a base, so the 13- to

23-face foldings may be reduced with the 12-face folding as a base.

The straight model which is the base for this process is colored gray where layered parts occur, and opening out these areas reciprocally yields the template for the desired n face folding. For more details refer to [5].

I was able to construct paper templates for all the models such that $9 \leq n \leq 24$, and by folding them confirm that they could all be produced in accordance with theory. I proceeded to complete the simple models first, and the 19-face model remained unresolved until the end. When $n = 19$ it is prime, and I was worried that this model might not be possible, but I settled on the positions by trial and error, and producing an expansion diagram revealed a snake-like form (Figure 9).

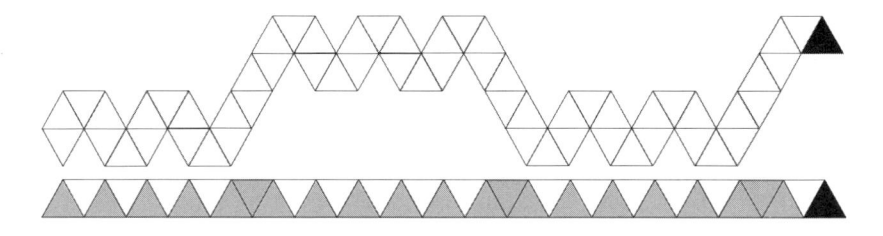

Figure 9: Template for the 19-Face Folding

It was demonstrated above that the cases when $3 \leq n \leq 24$ are possible, but this does not constitute mathematical proof for an arbitrary case. Diligently investigating the cases when $n \geq 25$ will probably not reveal any problems, but using actual materials to make the models and confirm their construction is painful, and this may be thought of as the limit.

Drawing the templates using a ruler and compasses takes time and leads to errors. I therefore made a versatile model that may be applied to all the templates (Figure 10). This was achieved using about 30 lines of Visual Basic instructions. The paper template needed for an n-face folding may be cut out from this template using a pair of scissors.

First of all, it is impressive see the phenomenon of the hexaflexagon with the 3-face folding. One may next wonder if 4-face folding is possible. Seeing that 4-face folding is possible, one may wonder whether 5-, 6- and arbitrary n-face foldings are possible. This thought process is similar to the methods of *extension, generalization, continuity* and *equivalence* used in mathematics. The fundamental 3-face folding is quite impressive by itself, and I'd be very pleased if those readers who have not experienced this puzzle would try it and see.

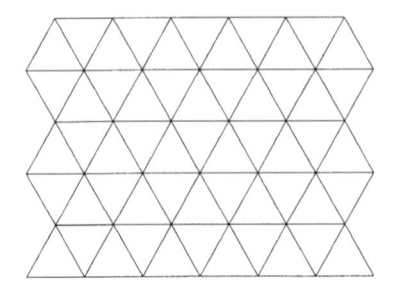

Figure 10: Versatile Template Made with Visual Basic

References

[1] M. Gardner, (translated by Y. Kanazawa), Origami Rokukakukei [Origami Hexagons], *The Scientific American Book of Mathematical Puzzles and Diversions*, Tokyo: Hakuyosha, (1960), 13-28.

[2] S. Ikeno, Tatamikae Origami [Layered Origami], *Puzzles IV*, Tokyo: Saiensusha, (1979), 78-82.

[3] J.S. Madachy, *Madachy's Mathematical Recreations*, Dover (1979).

[4] Y. Nishiyama, Origami Rokakukei [Folding Paper Hexaflexagons], *Basic Sugaku [Basic Mathematics]*, 23(12), (1990), 82-84.

[5] Y. Nishiyama, Hexaflexagons no Ippankai [General Solution for Hexaflexagons], *Journal of Osaka University of Economics*, 54(4), (2003), 153-173.

CHAPTER **11**
Turning Things Inside Out

Abstract: This article presents 3 puzzles which turn things inside out: the cube puzzle, turning a paper strip inside out and hexaflexagons. Readers who wish to find the solutions for themselves are advised not to read the full explanations.

AMS Subject Classification: 51H02, 00A09, 97A20
Key Words: Cube puzzle, Hexaflexagon, Möbius strip, Turning things inside out

1. Cube Puzzle

I spent a year conducting research overseas at Cambridge University in England from April 2005. While enrolled as a visiting fellow at Saint Edmund's College I worked in a research laboratory in the university department known as the Centre for Mathematical Sciences. This is like an amalgamation of what would be the physics department and the mathematics department in the science faculty at a Japanese university, and Stephen Hawking has his office in this very building.

Figure 1: Cube Puzzle

I had never before experienced such a period of stay overseas, and I was interested in everything there was to see. Also, having a hopeless sense of direction, I often got my routes mixed up, took the wrong bus or guessing incorrectly set off in the wrong direction, but after a month I eventually settled down. Once when I lost my way I thought to myself, "This is no good, I'd better get a map," and so I went into a tourist information office. There I found an interesting cube made of wood (Figure 1).

Figure 2: Folding up the Cube

Cambridge is a small university town with a population of 100,000, but with 31 colleges, various university departments and research buildings, the university's breadth from edge to edge is beyond walking. There are many structures that have existed since the middle ages such as King's College, Saint John's College, Round Church, Saint Mary's Church, the statue of Henry VIII, the Fitzwilliam Museum, the Mathematical bridge over the river Cam, the Bridge of Sighs and many more. There are not only students attending university, on non-working days it is bustling with tourists. Each of the colleges has a vast grass compound known as The Backs where students go strolling or may be found lounging about. I felt that time moved so slowly there! The cube had photographs of these scenic attractions attached to its faces. It did not simply have one picture attached to each of its faces. It could be folded up horizontally and vertically in various directions, and different photographs could me viewed one after the other. There were 6 square photographs, and 3 photographs formed by connecting two squares which is a total of 9 scenic images. The mechanics of its folding are interesting, so I bought one and took it back to my lodgings.

Nine scenic photographs is a strange number, but I soon understood the reason. Recalculating the sizes of the photographs yields $6 + 3 \times 2 = 12$ squares. The number of faces on a cube is 6, and doubling this yields 12. This is equal to the number of photographs. I realized that 12 seemed to be a number related to the construction of this puzzle, so I investigated what kind of structure was behind the cube. Firstly, the smaller sub-cubes, as basic units, had an edge length of 3.5 cm.

There are a total of 8 such sub-cubes, and they are packed together in 3 dimensions as $2 \times 2 \times 2 = 8$ cubes (Figure 2, Figure 3).

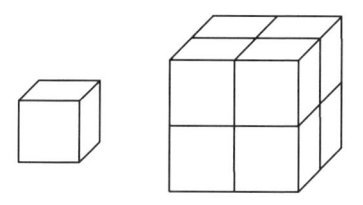

Figure 3: The 8 Sub-cubes

The wooden sub-cubes were made with unusual precision, and since they did not reveal a crack there was no sense of a join within the photographs. It was also made of a hard and heavy wooden material, so it made a satisfying 'clock clock' sound while it was being manipulated. The 8 sub-cubes were overlaid with what seemed like high quality film photographs. Maybe it was laminated - it didn't seem like it would be easily damaged.

Well, the most intriguing aspect is where and how the 8 sub-cubes are attached to each other. On the day that I bought it, I enjoyed analyzing its mechanism while manipulating it repeatedly. The scenic photographs were squares composed of four parts. As shown above there were a total of 12 faces covered with parts of these scenes. The basic unit is a sub-cube with 6 faces, and there were 8 such units, so the total number of faces is 48. Each scenic image is composed of sets of 4 faces, and dividing the 48 faces by 4 yields 12 sets. This means the puzzle is constructed in a way that makes expert use of all the sub-cube faces.

While manipulating it over and over again, I discovered that the 8 sub-cubes are attached at the places marked with thick lines in Figure 4. There are 8 places where the cubes are attached, and their positional relationship is interesting. If they are attached in this way the 8 sub-cubes do not fall apart separately, and all 48 of the faces can be revealed. I wonder who thought it up. A cube that can be turned inside out! It really is a wonderful idea.

Investigating this far made me want to take the next step and make one for myself. I thought about making a wooden model in the same way as the commercial item, but being abroad I couldn't easily lay hands on a saw and wood. I wondered whether it wouldn't be possible to make one out of drawing paper. I bought some paper, adhesive tape and glue

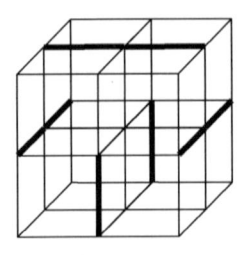

Figure 4: The Attachments between the Cubes

from a stationary shop, but I was surprised at how unusually expensive the price was. In Japan I was able to use paper in profuse quantities, but in England stationary was a precious commodity. I was further troubled by the complete absence of B series copy paper. In Japan, paper of all sizes - A4, A3, B4, B5 and so on - is readily available, but because Cambridge is a small city there was no B series to be found, only A series.

In the event, I made 8 cubes out of thick drawing paper, and while comparing them with the real thing, I sticky-taped them together. It took a while but I produced something resembling the original. It was a good thing I made it out of paper after all. Assembling it was a pleasure, and making it out of paper has the advantage that some degree of mismatch in size is permissible. Paper is soft, and the 8 sub-cubes kindly fit together harmoniously (Figure 5).

Figure 5: Turning the Home-made Cube Inside Out

The attachments are shown in Figure 4, and 24 of the faces of the small unit cubes are connected as a single surface. Let's take this as the outer surface. The 24 hidden faces (the inner surface) are also connected as a single surface. The 24 faces of the outer surface and the 24 faces of the inner surface are connected too, so the 48 faces of the whole body are thus connected. It can be manipulated in such a way that it is turned completely inside out, revealing that this is a cube reversal puzzle which involves turning a cube inside out.

For such a cube reversal puzzle there are surely only $2 \times 2 \times 2 = 8$ possible cases. For $3 \times 3 \times 3 = 27$, the manipulation would be impossible, so the number of faces would not match up. This aspect is an interesting mathematical point. Interested readers please make one for yourself.

2. Turning a Strip Inside Out

Talking about turning things inside out, I previously investigated another puzzle which involves turning things inside out in a different sense, and I'd like to introduce it here. The puzzle also involves cubes, but they are not the kind of cubes that are filled out completely like that in Figure 1, rather, it is a puzzle which involves connecting a paper strip into a cubic form. By making a model like that shown in Figure 6(a), and manipulating it repeatedly, it can be turned inside out into the form shown in Figure 6(b). The manipulation process is altogether like looking at a chienowa puzzle ring.

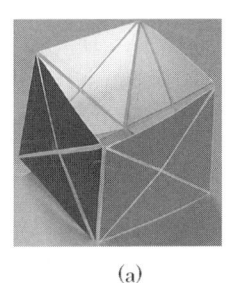

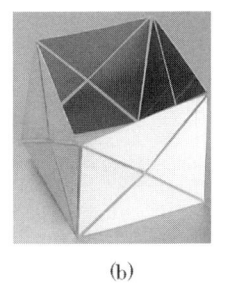

(a)   (b)

Figure 6: Turning the strip inside out

The real thrill of a puzzle is not looking at the solution, but rather the experience of solving it oneself, so I'll withhold the solution here. I'd like to encourage you to make this puzzle yourself. At the present time I have discovered two solution methods for turning it inside out. The fact that I am deliberately not supplying these solutions embodies my hope that a third solution might be found!

Now then, regarding how to make it, there is one method which involves making it from drawing paper and subsequently applying color, but I don't suggest this. The reason is that applying color using magic markers and so on may cause surface irregularities yielding a poor end result. Rather, art stores sell different types of art paper and it's better to buy these to use. It's best to choose your preferred colors from among clearly different colors of blue, green, red, *etc.* Then, taking care not to damage the colored side, measures can be marked on the white reverse side to guide construction. As shown in Figure 7, 4 squares with an edge length of 7 cm are connected horizontally. Diagonal lines are drawn to produce 16 right-angled isosceles triangles.

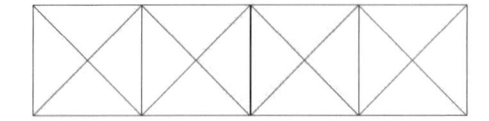

Figure 7: Construction Diagram for the Strip Puzzle

Hasty people may think that this completes the puzzle, but this is not the case. This crucial part is what follows. The 16 triangles are neatly cut apart with a pair of scissors. Next, the triangles are attached using sticky tape, with a gap between the triangles as shown in Figure 8, so that it can be operated smoothly. Since the edge of each square is 7 cm, a gap of about 2 mm should be suitable. If this gap is too large the visual appearance suffers, but if it is too small it cannot be easily manipulated.

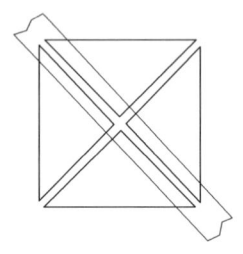

Figure 8: Sticking It Together with Gaps Using Sticky Tape

It's best to apply the sticky tape to the white side first, and if it goes well, then apply it to the colored side. This is because the colored side will not tolerate a mistake. The sticky tape is applied to both sides, extending beyond the edge of the squares. After applying the sticky tape to every part, the bits which stick out can be cut away. That is to say, I don't recommend reapplying or applying multiple layers of sticky tape. In this way, 4 squares connected by the sticky tape are produced. Connecting these horizontally into a strip completes the puzzle.

There is pleasure in solving this puzzle oneself. One may solve it while trying this and that, or alternatively find oneself stuck. There is definitely a route to a solution, and it must not be forced inside out. It is a peculiar thing, but knowing the route, the puzzle can be turned inside out easily. The interesting thing about this puzzle is that it is a strip which forms a cube. Increasing the number of faces in the strip to 5 or 6 faces, for example, would allow it to be turned inside out easily and it would cease to be a puzzle. The fact that the key to the puzzle is that there are 4 faces in the band which form a cube is strangely similar to the fact that the cube reversal puzzle shown in Figure 1 involves a cube.

For those readers who simply have to know a solution, there is Gardner's orthodox method [1]. The procedure has 12 steps and is a little on the long side, but it is a well-known process for turning the band inside out. Another method was taught to me by an acquaintance, and appears in *Mathematics Seminar* [2]. This method involves around 7 steps and is significantly shorter. It's a method for turning the band inside out that is brilliant enough to make one think, "Wow! This manipulation is possible!" I couldn't turn it inside out until I heard the solution.

Some people may find that they just can't turn the band inside out and seek a hint. In such cases I always say something like the following. Puzzles have a front and rear. This means that there must be a mid-point. The mid-point is when half the outer surface and half the inner surface are visible. If the mid-point is found, it is possible to get to the outer or the inner surface. This is just like climbing a mountain, where the mid-point is the peak, from which it is possible to return, or to descend to the other side. This means that if one can just find the mid-point, it is the same as having solved the problem.

3. Hexaflexagons

I have dealt with turning things inside out twice now, but the following hexaflexagon is also, in a sense, a puzzle which involves turning things inside out. The 'hexa' in hexaflexagon means 6, and the 'flexagon' refers to something which is flexible, *i.e.*, something that is possible to fold up. Figure 9 shows a hexaflexagon that I made myself. It is hexagonal, and as shown in Figure 10(a), by pinching two pairs of triangles, each in a diamond shape, a new face may be revealed from the center. Opening it out completely reveals a different face, as shown in Figure 10(b). There are 3 faces that can be revealed, and technically speaking it is not turned inside out, but rather it is in the same class of puzzles as the other two because they all involve revealing different faces.

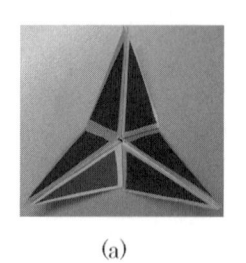

(a)

(b)

Figure 9: Hexaflexagon

Figure 10: Pinching in a diamond shape reveals a new face

The number of faces is 3 and, for example, by applying colors, blue, yellow and red faces can be revealed alternately. This is another peculiar puzzle, and for those readers who have never tried it, I'd like to explain how to make it.

Ten equilateral triangles with an edge length of 6 cm are lined up side by side as shown in Figure 11(a). The triangle on the right-hand end marked 'glue' is not related to the puzzle, it's just for gluing the hexaflexagon together. A diagram like this can be made with a pair of compasses and a ruler, so please draw it up for yourself. For the paper, standard copy paper is better than drawing paper. Copy paper is easy to fold and hard to tear.

The folding process begins by folding towards oneself along a-b as shown in Figure 11(b). Next a valley fold is made along c-d, leaving the part marked 'glue' exposed (Figure 11(c)). Then it is folded towards oneself again along e-f, and completed by gluing.

A Möbius strip is a normal strip glued into a loop with a 180° twist, and the hexaflexagon is a strip glued with three 180° twists, *i.e.*, a 540° twist. Gluing together with an odd number of twists results in a surface for which the inside and outside cannot be discriminated. This puzzle is an application of topology.

Once the puzzle is made, be sure to try manipulating it. If two pairs of triangles are pinched into diamond shapes, as shown in Figure 10(a), a new surface is unexpectedly revealed from the center. This is surely a delight that can only be experienced by people who make and manipulate the hexaflexagon for themselves. This puzzle is not exactly

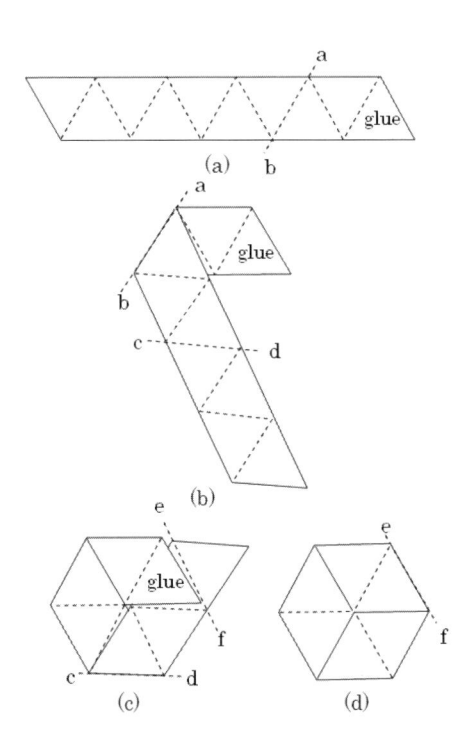

Figure 11: Three-face Folding Template and Folding Method

a 'turning inside out' puzzle, but it is in the same class in the sense that different faces are revealed.

After completing the 3-face-folding hexaflexagon in which 3 faces may be revealed, it's also fun to attempt the challenge of the 4-face-folding hexaflexagon. If you think about what kind of paper template and what kind of folding are needed to make the 4-face-folding model, I think you'll enjoy this puzzle too. After achieving the 4-face-folding, and while attempting a 5-face-folding, a kind of principle may be found, leading from the world of puzzles to the realm of mathematics.

I completed templates for the 4-face-folding up to the 8-face-folding model by referring to templates appearing in Gardner's book. I learned that the templates and folding methods for the 3-face, 6-face, 12-face-folding, *etc.*, and in general 3×2^n face-foldings, follow a principle. My interest was captured and I went as far as making a 24-face-folding model. I also learned the method behind the templates and folding patterns for the other numbers of faces. It was really quite an interesting experience. The topic is fully introduced in under the title 'A General Solution for Multiple Foldings of Hexaflexagons' in this book. Interested readers please take a look. However, this article includes the solution, so those readers who wish to find the solution for themselves are advised not to read it yet!

References

[1] M. Gardner, (trans. Y. Kanazawa), *Atarashi Sugaku Gemu Pazuru* [*The 2nd Scientific American Book of Mathematical Puzzles and Diversions*], Tokyo: Hakuyosha, (1968).

[2] Y. Nishiyama, Tukutemiyo Toitemiyo [Let's Make it, Let's Solve it!], *Sugaku Semina* [*Mathematics Seminar*], 33(10), 94-95, (1994).

CHAPTER 12
Miura Folding: Applying Origami to Space Exploration

Abstract: Miura folding is famous all over the world. It is an element of the ancient Japanese tradition of *origami* and its applications reach as far as astronautical engineering and the construction of *solar panels*. This article explains how to achieve the Miura folding, and describes its application to maps. The author also suggests in this context that nature may abhor the right angle, according to observation of the wing base of a dragonfly.

AMS Subject Classification: 51M05, 00A09, 97A20
Key Words: Miura folding, Solar panels, Similar diagrams, A and B paper size standards

1. An Application Involving Solar Panels

Perhaps you are familiar with the concept of Miura folding? Miura folding has a broad meaning, and is an element of *origami*. Its applications span from ancient Japanese traditions to astronautical engineering, and since it is also of interest mathematically I'd like to introduce it here.

The 'Miura' in Miura folding is derived from the name of the man who devised it, Koryo Miura. It was some time ago, but it is understood that this folding method occurred to him when he was researching aerospace structures while enrolled at Tokyo University's Institute of Space and Aeronautical Science. Rockets launched into space make use of the Sun's energy while they fly. The devices that gather this solar energy are solar panels, but these panels cannot be opened until after the launch. The solar panels are folded down as much as possible in order to pack them into the rocket, and then after the rocket blasts into outer space they quickly unfurl. When the rocket returns to the surface of the Earth, they must be folded down and re-stowed. The idea of Miura folding was realized after thinking about how this sequence of actions could be achieved not by humans, but by robots.

Miura folding has been introduced in newspapers and magazines and so on, and detailed discussions by the authors can be found in these articles. Here I would like for everyone to experience the wonder of mathematics by performing practical experiments on the principle of Miura folding [1-3].

Utilizing Mirua folding in this case involves holding the bottom left and top right corners of a piece of paper with the fingers, as shown in Figure 1. Placing the paper on a desk, a single movement can be used to open and close the paper if it is pulled in diagonally opposing directions. The fold behaves as if it has been remembered, so it can be described as shape-memory *origami*. This can't be done with usual maps. According to *origami* specialists, Miura apparently did not discover this type of fold himself, but recognizing the hint from ancient Japanese *origami* and applying it to solar panels was a great achievement.

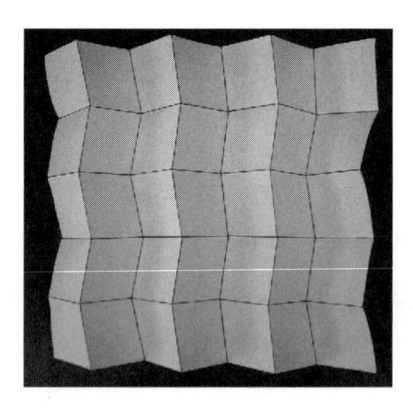

Figure 1: Miura Folding

2. Let's Try a Miura Folding

Allow me to explain the Miura folding method following the description by Koryo Miura himself in the book 'Solar Sails' [3]. I'd like the reader to attempt the folding according to Figure 2.

(1) First prepare a piece of A3 paper. B4 may also be used, but as the folds build up it gets smaller and smaller. The larger the paper, the easier it is to fold, and I therefore recommend A3.

(2) Fold the paper into 5 equal vertical parts, and 7 horizontal parts. If A3 paper is used then the vertical height is 297 mm so it cannot be exactly divided into 5 parts. Since it doesn't matter whether the two ends are too short, or too long, for the time being the part right in the middle should be maintained at the same length. The five vertical divisions should alternate mountain folds and valley folds, like a concertina.

(3) Next, it is folded horizontally into 7 divisions. The objective is to apply the diagonal fold shown by the dotted line. When 7 horizontal layers have been made, the 3rd layer from the left is bent around diagonally. The diagonal fold is made such that the tips are at a ratio of about 2 to 1. Such a steep angle makes it easy to achieve the Miura fold.

(4) Next, the 1st layer is folded back along its length. At this point the initial horizontal line and the current horizontal line should be made parallel.

(5) The diagonal is bent back in the same way. Again this should be parallel to the original diagonally folded line, *i.e.*, a zig-zag should be repeated. The left and right edges of the folds should be layered exactly. Only the last tab of paper is not layered.

(6) Flip it vertically in this state. The reverse side should be folded up in the same way using a repeated zig-zag while keeping each part parallel.

(7) The first half of the Miura folding is now complete. Some people mistakenly believe that this is the Miura folding itself, but in fact this is no more than a preparation for Miura folding.

(8) At this point, just once, let's spread out the paper on a desk and take a look. It has 5 vertical divisions, and 7 horizontal divisions. The horizontal lines are parallel, but the vertical lines are zig-zagging diagonals. Perhaps you will notice that the smallest element is a parallelogram. This is an essential condition of Miura folding.

The paper is folded vertically from the edge on the left using a mountain fold. The next edge is folded with a valley fold. I think you will notice as you fold it and see, but since the whole shape is connected in Miura folding, it cannot be achieved by repeating only individual mountain folds. Thus, with the paper gripped between the finger tips, we can

CHAPTER 12

(1) Prepare a sheet of A3 or B4.

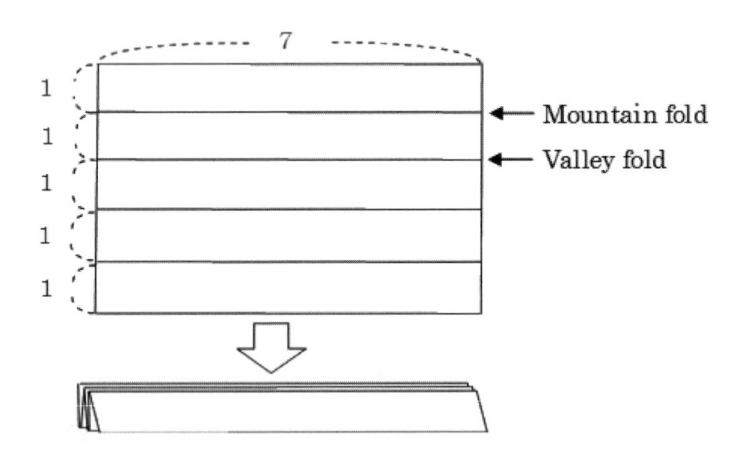

(2) Fold it into five sections vertically.

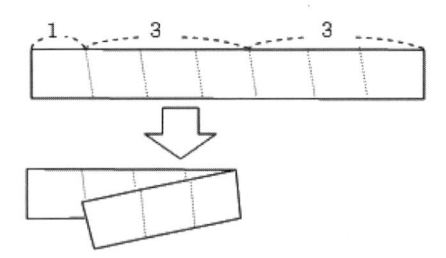

(3) Fold the length of the 3rd layer diagonally.

······· Parallel

(4) Fold back the length of the 1ˢᵗ layer parallel to the initial line.

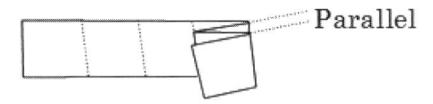

······· Parallel

(5) Make a repeated zig-zag.

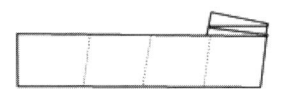

(6) Flip it vertically and fold the reverse side in the same way.

(7) At this point half of the Miura folding is now finished

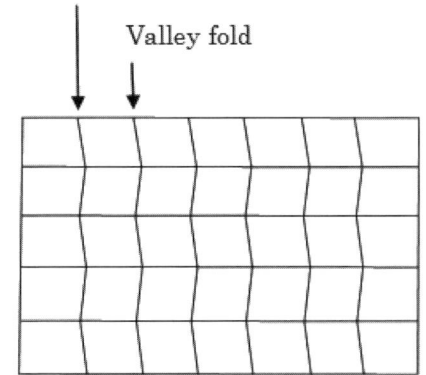

Mountain fold   Valley fold

(8) Spread the paper out on a desk.

CHAPTER 12

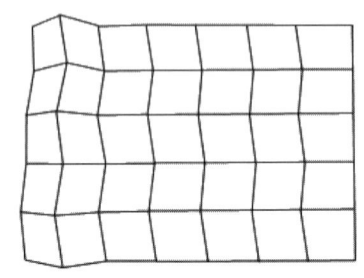

(9) Fold the leftmost column with a mountain fold, and then the next column with a valley fold.

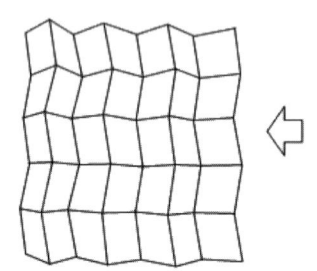

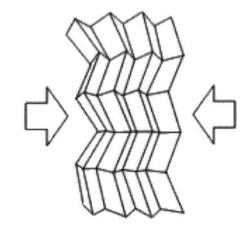

(10) Keep repeating the sequence mountain fold, valley fold, mountain fold, valley fold.

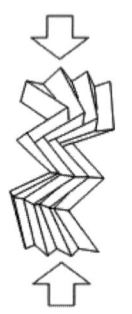

(11) Fold down the left and right.

(12) Fold down the top and bottom at the same time.

(13) The completed Miura folding

Figure 2: The Miura Folding Procedure [3]

stop without absolutely completing each fold. Also, if the valley folds are difficult, flipping the paper over and using a mountain fold might be easier.

By repeating the mountain folds and valley folds following steps (9) to (13), the whole body is collapsed down to the left hand side. Since the whole body is connected in Miura folding, collapsing down the left-right axis also collapses the vertical axis at the same time. In order to avoid damaging the folds during this process, it is important to proceed carefully. In this way, Miura folding is completed. Everyone should confirm that the completed Miura folding can be opened and closed in a single motion like that shown in Figure 1, by pinching it between the fingers, in the bottom left and the top right corners.

3. Similar Diagrams

A3 or B4 paper was used for the Miura folding with 5 vertical divisions and 7 horizontal divisions. Allow me to explain now why this size of paper and odd number of divisions were used.

According to JIS standards, paper sizes may be one of two types, A series and B series. The area of a piece of A0 is $1m^2$. Half this size is A1, taking half again yields A2, and so on for the A series. The area of a piece of B0 is $1.5m^2$. Half this size is B1, and half again is B2, and so on for the B series. The dimensions of the A series and the B series from 0 to 6 are shown in Table 1. The units are millimeters.

The interesting thing from a mathematical perspective is that the A and B series are both composed of similar diagrams. Since they are similar, the ratio of the vertical to the horizontal is the same for all the shapes, which are rectangles. The fact that copy paper sizes are similar diagrams should have been learned in junior high-school, but after finishing their exams, many university students and members of society completely forget about this fact. It's a shame that when you ask them to obtain the ratio of the vertical and horizontal dimensions of copy paper, most people end up unable to answer. But rather than committing Table 1 to memory, I'd like for the reader to understand the principle of similarity, and be able to assemble an equation and obtain a solution in this way.

It's possible to find the ratio of the vertical and horizontal dimensions in the following way. Let's denote the ratio of the rectangle's vertical and horizontal dimensions as 1 to x. Thinking about half of this rectangle, its vertical dimension will be $\frac{x}{2}$, and its horizontal dimension is 1, so the ratio is $1 : x = \frac{x}{2} : 1$. Solving this equation yields $x = \sqrt{2}$, *i.e.*, the ratio of the vertical and horizontal dimensions of copy paper is $1 : \sqrt{2}$, where $\sqrt{2} \approx 1.4142$.

The divisions used in Miura folding are 5 vertical and 7 horizontal divisions. Since the ratio of the vertical and horizontal dimensions of the largest element is $\frac{7}{5} = 1.4$, this value is close to $\sqrt{2}$. This means we can expect the benefit that the element can be folded up with a shape close to a square. Also, both 5 and 7 are odd numbers of divisions. If the number of divisions is odd, then when the paper is gripped between the fingertips in the bottom left and top right corners and pulled, the paper does not flip over, but rather it spreads out. Any number of divisions in the vertical and horizontal directions should be acceptable for Miura folding, although the reference above has taken care to investigate all the possible configurations in this neighborhood.

A series	No.	B series
841 × 1189	0	1030 × 1456
594 × 841	1	728 × 1030
420 × 594	2	515 × 728
297 × 420	3	364 × 515
210 × 297	4	257 × 364
148 × 210	5	182 × 257
105 × 148	6	128 × 182

Table 1. JIS Standard Paper Sizes (mm)

4. Application to Maps

Solar panels which apply the principle of Miura folding have actually been loaded onto the experimental Japanese satellite N2, and spread out in space. Owing to the principle of Miura folding they were spread in a single motion, but I heard that the closing motion in order to pack them away did not proceed well. Perhaps it is harder to close it than to open it.

Miura folding is truly wonderful. When I introduced it at a symposium on mathematics education, one of the participants informed me that they had discovered a map that utilized Miura folding. It was being sold by the Kyoto tourist board. I quickly made arrangements and ordered one. The map of Kyoto city center was indeed made using Miura folding. Perhaps tourists might take out the map from an inner pocket, spread it open with a single movement, confirm their destination, then close it once again with a single movement and put it back in their pocket. Besides the Kyoto city center tourist map, there was also a road map of the highways in the capital. However, the recent advancements in car navigation systems might spell the gradual disappearance of traditional paper road maps.

Well, now you've persevered with my review of Miura folding using 5 vertical and 7 horizontal divisions as explained above, let's move on. The point behind Miura folding is that the horizontal lines are parallel while the vertical lines are in a zig-zag. If the verticals and horizontals are both parallel, that is to say the vertical and horizontal lines are at right angles, then it cannot be opened and closed with a single movement in the manner of Miura folding. I'd like for the reader to make a model with 5 and 7 divisions using normal folding, and then perform a comparative investigation with Miura folding.

Also, the folds in maps made using Miura folding are slightly offset, with the result that they are difficult to cut like normal maps. Refer to Figure 2(13). Miura folding has parallel horizontal lines, and zig-zagging vertical lines. I wondered if it could be made with zig-zagging horizontal lines as well. This is interesting mathematically, and is possible. Attempting to confirm this by drawing up diagrams revealed that it could be folded just by allowing the positions of the parallelograms to be irregular [4]. However, I'm not sure how meaningful this really is.

5. Does Nature Abhor the Right Angle?

It is said that Koryo Miura devised Miura folding in 1970 after observing the wrinkles in old people's brows and in the surface of the Earth in photographs taken from spaceships. The idea behind Miura folding was obtained through a detailed observation of nature.

The long running author Toda Morikazu of the 'Toys Seminar' in the periodical *Mathematics Seminar*, has also dealt with Miura folding [5]. It is in the section entitled 'Snakes on the Move.' This article deals with toy 'paper snakes' and explains the mechanism of a cornice. It is taken that since snakes advance by extending and contracting, they must bend themselves in a zig-zag similar to Miura folding.

Thinking along those lines, the cornices in Chinese lanterns and cameras all zig-zag in the same way. Isn't it true that right angles are no good for folding up nicely like this? It is thought that the blood vessels in the bases of dragonfly and butterfly wings are not orthogonal. Perhaps when resting with the wings closed, right angles would prevent the wings from being neatly folded away. Figure 3 shows the base of Cordulegasteridae wings [6]. The anterior edge of the wing base is zig-zagged like Miura folding. The pattern of blood vessels is also parallel, and it is complex with few right-angled components visible. The dragonfly develops from a larva, metamorphoses into an adult insect, and the wings open from a closed condition, so there is some relationship with Miura folding. The straight lines we learn about in mathematics, circles, 2nd order functions, as well as curves and so on are simple because they are artificial. These kinds of curves rarely exist in the natural world. Perhaps there's a reason for the complex patterns? The progression up to the present day must certainly have required a great many years.

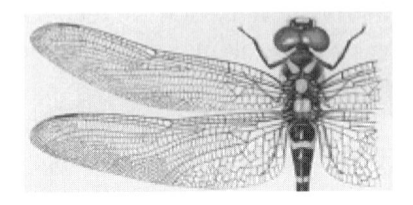

Figure 3: The Wing Base of a Dragonfly [6]

References

[1] Asahi Newspaper, Hito, Miura Koryo / Miuraori te Nandesuka [People, Koryo Miura / 'What is Miura folding?'], (Nov. 30th 1994).

[2] K. Miura, Utyuni hiraku Mahono Origami [The Magic "Origami" that Opens in Space], *Kagaku Asahi*, (Feb. 1988).

[3] K. Miura, N. Nagatomo, *Sora Seiru [Solar Sails]*, Tokyo: Maruzen (1993).

[4] Y. Nishiyama, Miuraori wo Tukutemiyo [Let's Try Using 'Miura folding'], *Sugaku Kyoshitu [Mathematics Classroom]*, 41(6), 93-96, (1995).

[5] M. Toda, *Zoku Omotya Semina [Continuing Toy Seminar]*, Tokyo: Nihon Hyoronsha, (1979).

[6] Shogakukan, *Kontyuno Zukan [Picture Book of Creepy Crawlies]*, Tokyo: Shogakukan, (1987).

CHAPTER 13
The *Sepak Takraw* Ball Puzzle

Abstract: This article explains the truncated icosahedron by making a *sepak takraw* ball, which is used in a popular ball game in Thailand. Next, many popular topics for examination questions on spatial diagrams are presented such as: how many regular polyhedra can be made, Euler's polyhedron formula, $F - E + V = 2$, and the number of regular pentagons and hexagons in a soccer ball.

AMS Subject Classification: 51M05, 00A09, 97A20
Key Words: Truncated icosahedron, Regular polyhedra, Euler's polyhedron formula, Fullerene C_{60}

1. Let's Make a *Sepak Takraw* Ball

Both stress and relaxation are important when studying for an examination. After the challenge of university entrance examinations it makes sense to find some enjoyment in puzzles. The diversion I'd like to introduce was revealed to me by Shigenori Ohsawa, a high-school teacher in Saitama prefecture. In Thailand there is a popular ball game called *sepak takraw*. Apparently, '*sepak*' means 'kick' in Malay, and '*takraw*' means ball in Thai. The ball used in this competitive sport is very durable. It is made of rattan and is 12 cm in diameter. The game involves kicking the ball, and is similar to the Japanese sport of

Figure 1: *Sepak Takraw*

kemari (Figure 1). This has become a medal sport at the Asian Games and uses a plastic ball.

The *sepak takraw* ball is related mathematically to a 32-face semi-regular polyhedron, known as a truncated icosahedron. In fact, it is not merely engaging as a plaything, but is also related to the fullerene C_{60} molecule made famous by the award of a Nobel Prize in chemistry, as well as the construction of soccer balls (Figures 2 and 3).

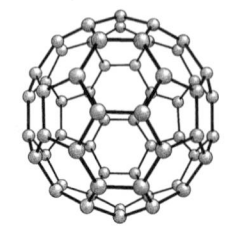

Figure 2: Fullerene

Figure 3: Soccer Ball

What I'd like to introduce here is how we can construct a ball with the same form as the *sepak takraw* ball using simple packing tape. By way of preparation, a 3-4 m length of packing tape (made from polypropylene, known as PP-band) 15 mm wide will be sufficient. The tape should be cut into 6 pieces, 56 cm in length. This length and number of strips will be explained later. It's also useful to have 6 clothes pegs which can be used to hold things in place while working.

The following four points regarding the construction are important.

(1) Each strip winds around two 25 cm loops with 6 cm to spare. Two loops are used in order to increase the strength.

(2) Three-way locks are formed where three strips cross (Figure 4). There is another such pattern which is the other way up, but the patterns are equivalent.

(3) When tapes cross each another, they alternate crossing above and below.

(4) Five strips of tape make a regular pentagonal hole (the black part in Figure 5).

It can be seen that in the area indicated with a dotted line, the tape forms a regular hexagon. Keeping the above-mentioned basics in mind, allow me to explain the construction procedure with reference to Figure 6. First there is the assembly of the three-way lock (1). This relationship occurs at every crossing point so as a fundamental, it should be memorized.

Next is the construction of the regular pentagonal hole from five strips of tape (2). When clothes pegs are used to hold the tape in place, the process is easy when the pegs are arranged on radial lines. The number of clothes pegs is five. The point to remember is to pass the

SEPAK TAKRAW **139**

Figure 4: Three-way Lock

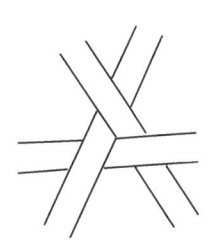

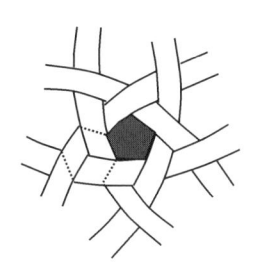

Figure 5: Making a Regular Pentagon

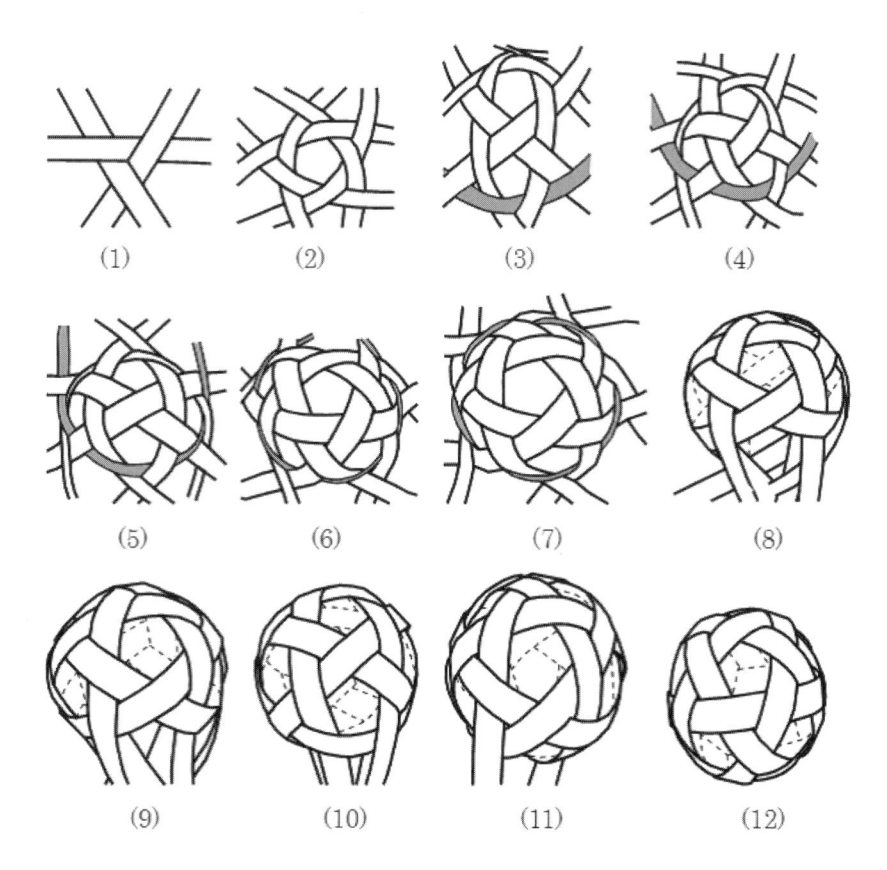

Figure 6: Construction Diagram for a *Sepak Takraw* Ball

sixth strip through them. It doesn't matter where from, but it is best not to forget the three-way lock and pentagonal hole construction. The sixth strip is shown in gray. The second regular pentagon is made (3), then the third (4), then the fourth (5), the fifth (6) and the sixth (7).

At this point the arrangement of the six regular pentagons is such that there are five around the original pentagon. Also, the sixth tape (gray) is a closed loop when the six pentagons are completed, and is wrapped twice.

The whole object is now half complete, and the clothes pegs used we no longer need the clothes pegs to hold it together (7). The *sepak takraw* ball is completed by passing the five free strips through in order, above and below (8-12).

2. How Many Regular Polyhedra Can We Make?

The *sepak takraw* ball does not involve solving mathematics with pencil and paper; instead it is constructed by hand and thus gives us a sense of the physical realization of mathematics as well as helping us to understand spatial diagrams. There are many problems that may be presented as examination questions on spatial diagrams and I'd like to introduce a few of them here.

Among polyhedra, there are regular polyhedra and semi-regular polyhedra. When the elements in the construction of a polyhedron, the polygonal faces, are all one type of regular polygon, the polyhedron itself is described as regular. When there are two or more types of regular polygon, the polyhedron is described as semi-regular. It is known that there are five regular polyhedra as shown in Figure 7: the tetrahedron (4 faces), cube (6 faces), octahedron (8 faces), dodecahedron (12 faces) and icosahedron (20 faces). These five regular polyhedra thus exist, and it has been proven that there are no others. Perhaps this is something to learn by heart? I don't think so, because with just a little thought this fact can be easily proved. So let's take a look [2][3].

[Problem 1] Prove that there are only five regular polyhedra.

(Proof) Let's try to prove it by thinking of a regular polyhedron as an object in which m regular n sided polygons meet at a single vertex. The angle at a single vertex of a regular n sided polygon is $180° - \dfrac{360°}{n}$. Supposing m regular n sided polygons meet at a vertex, then the total angle formed by all the individual vertex angles must be less than $360°$ so,

(a) Tetrahedron (4 faces) (b) Cube (6 faces) (c) Octahedron (8 faces)

(d) Dodecahedron (12 faces) (e) Icosahedron (20 faces)

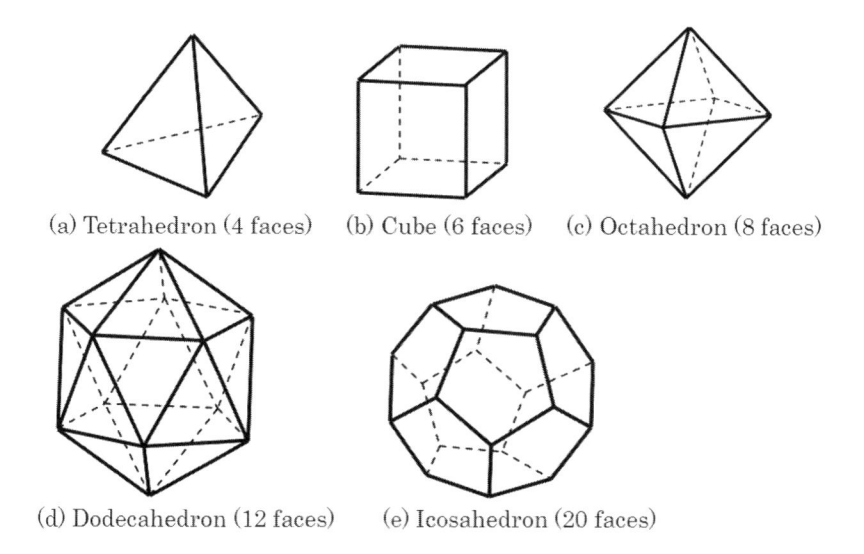

Figure 7: The Five Regular Polyhedra

$$(180° - \frac{360°}{n})m < 360°.$$

Rearranging this,

$$(m-2)(n-2) < 4 \quad (n, m \geq 3).$$

Solving this yields,
  for $n = 3$, $m = 3, 4, 5$ (tetrahedran, octahedron, icosahedron),
  for $n = 4$, $m = 3$ (cube),
  for $n = 5$, $m = 3$ (dodecahedron).   (End of proof)
Check each of the cases above!

3. Euler's Theory of Polyhedra

The tetrahedron is constructed from 4 triangles, the cube from 6 squares, the octahedron from 8 triangles, the dodecahedron from 12 pentagons and the icosahedron from 20 triangles. There is also the 32-face semi-regular polyhedron known as a truncated icosahedron which is constructed from 12 regular pentagons and 20 regular hexagons (totaling 32 faces).

Writing out the number of faces, edges and vertices of the regular polyhedra and this semi-regular polyhedron yields Table 1.

	Faces (F)	Edges (E)	Vertices (V)	$F - E + V$
Tetrahedron	4	6	4	2
Cube	6	12	8	2
Octahedron	8	12	6	2
Dodecahedron	12	30	20	2
Icosahedron	20	30	12	2
Truncated icosaderon	32	90	60	2

Table 1. $F - E + V = 2$

Now, regarding the relationship between these numbers, calculating (Faces - Edges + Vertices) reveals that the following relationship is satisfied.

$$F - E + V = 2$$

This is known as Euler's formula, and proofs may be found in many books. I'll state a method here drawn from a book close to hand, 'The History of Geometry,' by Kentaro Yano [6].

[Problem 2] Describe a method of proving Euler's formula, $F - E + V = 2$, for an arbitrary polyhedron such as that shown in the diagram.

Consider the object that results from removing one face of the polyhedron, such as ABC for example. By so doing, the number of vertices V and the number of edges E are unchanged, but the number of faces is reduced by 1, $F' = F - 1$. Thus, although the objective is to prove for the original polyhedron that

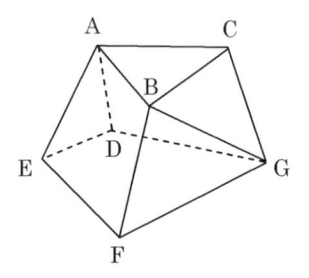

$$F - E + V = 2,$$

it is sufficient to prove that

$$F' - E + V = 1$$

for the object that results from removing one face in this way. The proof is made easier by partitioning the polygons into triangles. The faces adjacent to ABC are removed one by one, and it is sufficient to track how the value of $F' - E + V$ develops during this process. In the end, only one face is left (a triangle), and it is made clear that the relationship

$$F' - E + V = 1 - 3 + 3 = 1$$

is satisfied. (End of proof)
Refer to the above-mentioned book for details.

4. Fullerene Molecules and Truncated Icosahedra

Perhaps you know that the discovery of the fullerene C_{60} molecule was connected to Euler's formula [1]? At first, for this new molecule with 60 carbon atoms, chemists only considered regular hexagons for the distribution model of the atoms. However, it was understood that it is not possible to cover a spherical surface using such a method. In terms of chemical formulae, a spherical surface formed from six-member rings is not closed, so the existence of five-member rings is theoretically necessary. This issue has been presented as an examination problem.

[Problem 3] Use Euler's theory of polyhedra, $F - E + V = 2$, to prove that graphite sheets formed from 60 atoms cannot be bent and closed in order make a polyhedron.

Having 60 atoms means the number of vertices is 60 ($V = 60$). Since each vertex is shared by 3 polygons, each vertex joins 3 'half-edges.' This is because each edge is shared by two vertices. The total number of edges is therefore 90 ($60 \times 3 \div 2 = 90, E = 90$). Assuming all of the faces are hexagonal, since each edge accounts for one sixth of two faces (one on either side of the edge), the total number of faces is 30 ($90 \times 2 = 180, 180 \div 6 = 30$), ($F = 30$).

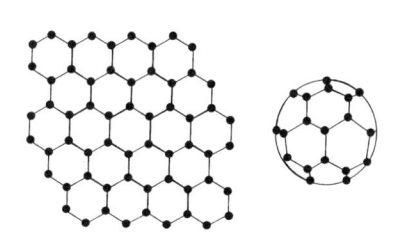

Substituting this value into the left hand side yields

$$F - E + V = 30 - 90 + 60 = 0,$$

for which the right-hand side is not equal to 2. It is therefore impossible to make a C_{60} polyhedron from a graphite sheet containing only hexagonal surfaces. (End of proof)

I have already mentioned that there are only five regular polyhedra in existence, but there are the semi-regular polyhedra which relax the conditions on the construction of regular polyhedra by permitting two or more

Figure 8: Truncated Icosahedron (32 Faces)

different types of regular polygon. A representative example of semi-regular polyhedra is the truncated icosahedron which has 32 faces, which is the closest to a sphere among all of the semi-regular polyhedra, and is also recognizable as a soccer ball!

The fullerene molecule is written C_{60}, and these 60 carbon atoms are arranged as the vertices of a polyhedron. The molecule is mathematically equivalent to a 32-face truncated icosahedron. Here, a '32-face polyhedron' expresses the number of faces, but the number of vertices is 60, and there are 90 edges, so Euler's formula takes the following value.

$$F - E + V = 32 - 90 + 60 = 2$$

5. The Number of Regular Pentagons and Hexagons

We know that the truncated icosahedron is constructed from 12 regular pentagons and 20 regular hexagons with a total of 32 faces, and I once presented a problem for readers of a math magazine to determine the number of pentagons and hexagons while examining a photograph of a soccer ball [4]. This is quite a nice problem, and I think it may be ideal for an entrance exam, so let's take a look at it now.

[Problem 4] A soccer ball is constructed from a total of 32 regular pentagons and hexagons. From the information that can be obtained by looking at the photograph alone, determine the number of pentagons and hexagons individually.

There have been many unique answers, so let's look at the orthodox solution. Denote the number of regular pentagons by x, and the number of regular hexagons as y. Since the total is 32,

$$x + y = 32.$$

The problem is how to introduce one more equation. Suitable equations focusing on the number of faces, the number of edges or the number of vertices are all possible.

[Focusing on faces]
Since there are 5 regular hexagons around each regular pentagon, if overlaps are permitted then the total number of hexagons would be $5x$. On the other hand, around each hexagon there are 3 pentagons, so $5x$ counts 3 overlapping layers. Thus, if $5x$ is divided by the degree of overlapping, 3, then it becomes the actual number of regular hexagons.

$$y = \frac{5x}{3}$$

Solving this yields $x = 12$ and $y = 20$.

[Focusing on edges]
The number of edges around a regular pentagon is 5, so there are $5x$ such edges in total. The number of edges around a regular hexagon is 6, so the total number of such edges is $6y$. Each regular hexagon is only adjacent to a regular pentagon along every other edge, so half the number of hexagon edges is equal to the number of edges around the regular pentagons.

$$5x = \frac{6y}{2}$$

Solving this yields $x = 12$ and $y = 20$.

[Focusing on vertices]
A regular pentagon has 5 vertices, so there are a total $5x$ such vertices. A regular hexagon has 6 vertices, so there are a total of $6y$ such vertices. Paying attention to a single vertex, there is 1 adjacent pentagon, and there are 2 adjacent hexagons. Since the number of hexagon vertices, $6y$, is twice the number of pentagon vertices, $5x$,

$$5x : 6y = 1 : 2,$$

Solving this yields $x = 12$ and $y = 20$.   (End of proof)

None of the solution methods above involves complicated equations, but it's clear that having ability for spatial interpretation is particularly important in order to derive these equations. The proportion of entrance examination questions involving mechanical calculation is growing large, but aren't problems that test mathematical thinking also necessary?

6. The Width of the Tape and the Radius of the Ball

After making several *sepak takraw* balls from PP-band tape, you begin to wonder just how big a ball you can make. I therefore looked into the relationship between the width and length of the tape, and the radius of the ball [5].

[Problem 5] When the width of tape used to make a *sepak takraw* ball is d, find the length, L, of the strip that wraps the around the ball once. Think about how a ball with double the radius would be made.

For the strip to encircle the ball once, it must pass through the 10 hexagons from A to B shown in Figure 9, so the length of the loop is $L = 10\sqrt{3}d$. The value of this coefficient is roughly 17.3. This relational expression may be referred to when estimating how much tape is needed. If the width of the tape is 15 mm, then the length of a loop is approximately 26 cm.

Taking another perspective, by considering that the tape must pass through the course around the truncated icosahedron as shown in Figure 10, the length of the loop L' is

$$L' = \{8 + \frac{4\sqrt{3}}{3}(1 + \sqrt{5 + 2\sqrt{5}})\}d.$$

The value of the coefficient in this expression is 17.4, and is slightly larger than the previous value.

Whichever is used, the length of the loop of tape may be expressed as a proportion of the width. The length of the loop and the radius of the ball are related in the proportion $L = 2\pi r$, so the radius of the ball is proportional to the width of the tape.

This means that if you want to produce a *sepak takraw* ball with twice the radius, you don't need strips of tape that are twice as long, but rather strips that are twice as wide.

It is common in high school mathematics not to deviate from textbooks and reference books. But how about starting afresh and studying mathematics by making things? It may feel like the long way round, but it might in fact be a surprising shortcut.

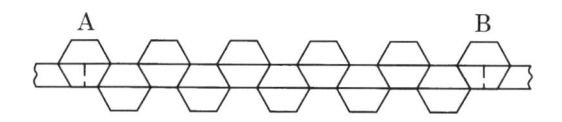

Figure 9: For the Strip to Encircle the Ball Once

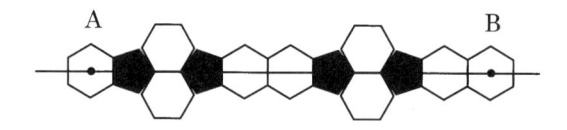

Figure 10: The Course around the Truncated Icosahedron

References

[1] J. Baggott, *Perfect Symmetry*, Oxford University Press, (1996).

[2] S. Hitotsumatsu, *Seitamentaiwo Toku* [*Solving Regular Polyhedra*], Tokyo: Tokai University Press, (1983).

[3] I. Murakami, *Utsukusi Tamentai* [*Beautiful Polyhedra*], Tokyo: Meiji Books, (1982).

[4] Y. Nishiyama, Elegant na Kaitowo Motomu [Seeking Elegant Solutions], *Sugaku Semina* [*Mathematics Seminar*], 33(9), 94-99, (1994).

[5] Y. Nishiyama, Sepatakuro Boru no Hankei [The Radius of a *Sepak Takraw* Ball], *Sugaku Kyositsu* [*Mathematics Classroom*], 47(12), 92-93, (2001).

[6] K. Yano, *Kikagakuno Rekishi* [*The History of Geometry*], Tokyo: NHK Books, (1972).

CHAPTER **14**

Increasing and Decreasing of Areas

Abstract: This article explains how to help junior-high school students enjoy studying geometry using some card magic. Area may be increased and decreased by arranging pieces of card according to the images on either the front or back of the card. There are some interesting stories associated with this puzzle, including 'Sunflowers Facing the Sun' and 'The UFO-Spotting Brothers.'

AMS Subject Classification: 51M05, 00A09, 97A20
Key Words: Area, Geometry

1. The Disappearing Square Piece

It's growing close to 20 years since I started doing this, but there's a mathematical puzzle I still like to introduce to students every year during seminar time. I am very familiar with the puzzle, so for me it lacks freshness, but students who come upon it for the first time seem impressed all the same. The puzzle that I'll now introduce presents a surprising phenomenon that also makes one want to try to make something similar oneself, as well as try to unlock the secret of the trick mathematically. As a teaching resource for mathematical education it thus kills three birds with just one stone.

First, allow me to introduce this puzzle I first came across in a magazine. The sliced up playing card on the desk shown in Figure 1(a) is face down. With the help of a student, have these pieces arranged so as to reform the pattern, just as if it were a jigsaw, as shown in Figure 1(b). There are 5 pieces in total: two trapezoids, a right-angled triangle, a rectangle and one square. Since there are only a few pieces, the reverse side is soon completed. After confirming that they are all back together, have the pieces mixed up like at the beginning (a), while keeping them face down.

Then, keeping the 5 pieces where they are, have them turned face up as shown in diagram (c). At this point it is clear that the card was the king of diamonds. Have the pieces arranged together just as before. The design on the front of the card is reformed, but the square piece is left over (d). The puzzle is therefore to work out how this piece can be left over.

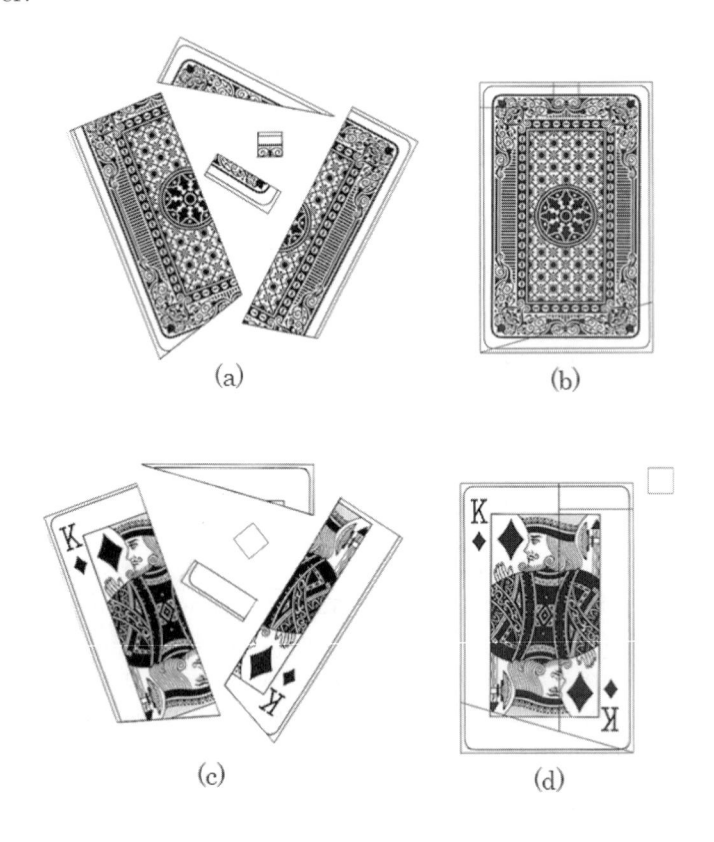

(a)

(b)

(c)

(d)

Figure 1: The Disappearing Square Piece

2. Making the Puzzle

When they see this phenomenon, most students are surprised. At that point I explain that it is neither an illusion nor magic, so there is no secret trickery or device involved, everything is open, and that I'd like them to think about an explanation. Next I have them try to make something similar. The students are given a print out resembling the design on a playing card as shown in Figure 2, and some card or thick drawing paper onto which they can glue the design. The image of the playing card used is a copy blown up to an appropriate size. A height of 15 cm with a width of 10 cm should be suitable.

(a) Back   (b) Front

Figure 2: A Playing Card for Constructing the Puzzle

Using the design from a playing card is very convenient during construction. First, place the image from the back side of a playing card on the card (Figure 3). Then mark the grid points on the vertical line with a compass needle, or by pressing down hard with a ball point pen. There are 11 grid points in total. After the marks have been made, remove the design and draw a line joining the marked points. The card is now ready (see the right-hand side of Figure 3).

The thin paper playing card shown in Figure 2 (front and back) used for the construction, and the thick card prepared according to Figure 3 are cut out using a pair of scissors. The number of pieces cut is $5 \times 2 + 4 = 14$. These pieces are to be glued onto the card, but since the thin paper is prone to stretch, it seems to be best to apply the glue to the card. The glued paper may also have a tendency to bend, so it should be kept flat. So now we have the playing card for the puzzle.

3. Thinking About the Explanation

Once the glue has dried and the card is ready to use, group the students into pairs, and have them perform the puzzle using an assistant in the

152   CHAPTER 14

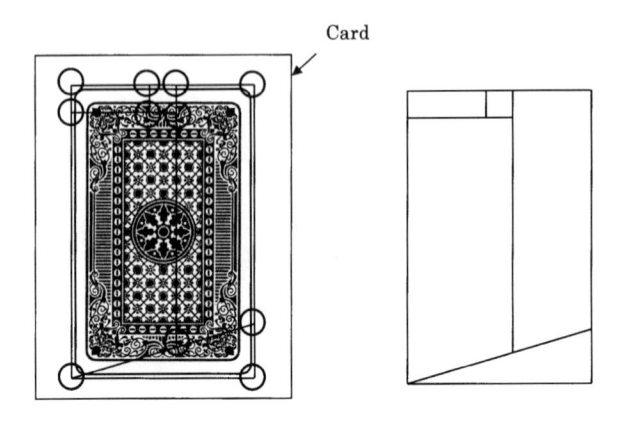

Figure 3: The 11 Grid Points

manner that I described earlier. One of the nice things about this puzzle is that anyone can be allowed to assemble the pattern, so any suspicion of hidden trickery or devices is dispelled.

After two or three attempts using trial-and-error, a student typically cries out "I've got it!" If the correctly assembled design on the back (Figure 1(b)) is flipped over without adjusting the places of the pieces, the design on the front is not assembled. This may be called an explanation. For me, this is saying no more than restating the phenomenon itself. The trapezoids on the left and right are certainly swapped. So why does swapping the pieces increase or decrease the area? I ask students to think about this, and at this point they become very quiet.

University students in the humanities enter university without doing much mathematics. After explaining that a knowledge of junior high-school geometry is sufficient for an understanding of this mathematical puzzle, I give a hint by a drawing diagram of the dimensions like that shown in Figure 4 on a whiteboard, in order to help students think.

These are the dimensions of the back of a playing card. It is sufficient to think about a playing card with a height of 10 cm and a width of 7 cm. The lengths of the bases of the two trapezoids are 9 cm and 8 cm. The height of the rectangle is 1 cm and its width is 3 cm. The base of the right-angled triangle is 7 cm and its height is 2 cm. There is an edge in Figure 4 which doesn't have a clearly marked length, and calculating it is the key to solving this problem. Thinking is important for puzzles, so I'll show an example calculation last of all. Please attempt this challenge yourself.

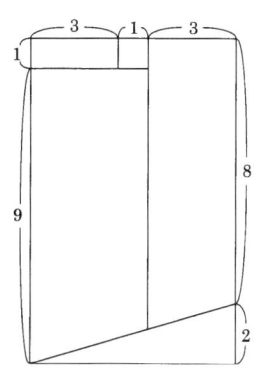

Figure 4: The Dimensions on the Back of the Card

4. 'Sunflowers Facing the Sun'

I first learned of this mathematical puzzle in 1987, which is now 20 years ago. I was watching an NHK television when some footage transmitted by the UK's BBC was broadcast. It was a program called 'The Paul Daniel's Magic Show.' Somehow, while I was watching the television I was completely drawn in by this puzzle. I remember hurriedly pressing the record button on the video and saving the footage.

Around that time, there was a page in the Sunday edition of the *Asahi* newspaper called 'The Natural History of Puzzles,' by Izuo Sakane, and this puzzle was taken up in this entertaining series of articles on toys and puzzles. In the *Asahi* newspaper they didn't use a design from a playing card, but instead a work by Hiroshi Kondo called 'Sunflowers Facing the Sun' [1]. This is shown in Figure 5, and has the same number of pieces as the playing card, *i.e.*, 5. There are sunflowers drawn on the two trapezoid pieces, and a sun on the square. On the left-hand side of Figure 5, while the sun is out the sunflowers are facing the sun, but when the sun sets (the square is removed) the sunflowers face where they like, as shown on the right-hand side.

The designs on playing cards have a sense of narrative, making the puzzle rather entertaining. Another design was also introduced by the teachers at a high-school mathematics education assembly. This was known as 'The UFO-Spotting Brothers.' The two friendly brothers are gazing at the night sky. The younger brother points and shouts "Look, a UFO!" Next the elder brother looks in the direction of the UFO, but the UFO has suddenly vanished. That is, when the piece with the UFO (the square) is removed, the brothers say to each other, "Where did the

Figure 5: 'Sunflowers facing the sun' by H. Kondo [1]

UFO go?" and search the sky in different directions.

Isn't this rather nice and romantic? How about devising your own story? Anyway, for both 'Sunflowers Facing the Sun' and 'The UFO-Spotting Brothers,' when the design is face up, the trapezoids on the left and right are exchanged. In the case of the playing card, turning over the whole thing reveals that the left and right parts are swapped. When solving this problem from a mathematical perspective all three cases are identical.

5. The Magic Conditions

So, has the hint regarding the dimensions shown in Figure 4 helped you realize the reason why the square piece has disappeared? The reason becomes apparent when the lengths of all the dimensions besides that hinted are calculated.

Speaking in terms of the result, the lengths concealed when the playing card is face down are $8\frac{6}{7}$ and $1\frac{1}{7}$ in the vertical direction (Figure 6(a)). Performing the calculation by focusing on the right-angled triangle, and using similar distances, $1\frac{1}{7}$ is first obtained. Junior high-school level geometry is sufficient, so everyone should confirm this for themselves. Summing these up, the result is 10, and the lengths on both the left and right ends are equal. Carefully exchanging the two trapezoids on the left and right produces the front surface of the playing card (Figure 6(b)). 'Carefully' here means taking care not to move the right-angled triangle and the rectangle, which are not repositioned.

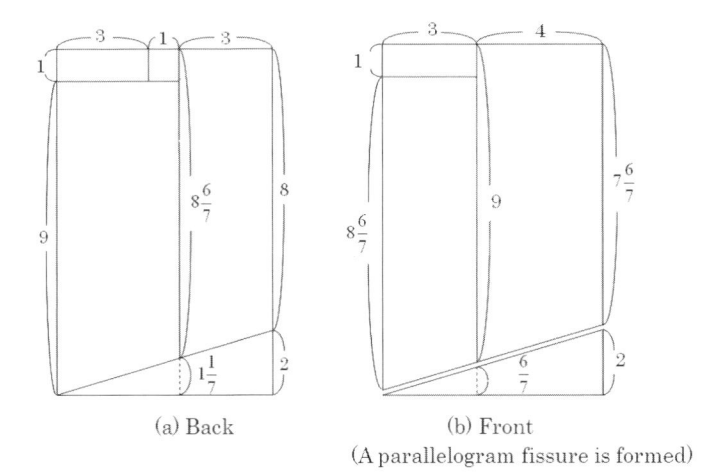

(a) Back

(b) Front
(A parallelogram fissure is formed)

Figure 6: All Dimensions of this Puzzle

By exchanging the trapezoids on the right and left, the width of 7 cm before the repositioning is not changed, but the height is slightly reduced. The 10 cm before the repositioning is slightly reduced to $9\frac{6}{7}$ cm. It's only $\frac{1}{7}$ cm, so no one notices. By the same token, the gradients of the right-angled triangle and the trapezoids are the same, so by trying to place these lines together, the fissure is naturally buried away.

Let's calculate the variation in area. The width and height of the square are 1 cm, so it has an area of $1 \times 1 = 1$ cm². Looking carefully at the fissure, it is a long thin parallelogram. Let's look at the parallelogram outlined in red, in order to emphasize the fissure. This reveals that the parallelogram really is long and thin, with a base of $\frac{1}{7}$ cm and a height of 7 cm. Since the area of a parallelogram is equal to its base width × height, the area is $\frac{1}{7} \times 7 = 1$ cm².

We can thus see that the areas of the square piece and the parallelogram are equal. The square piece was reshaped into the parallelogram. In the natural world there is a law dictating the conservation of energy, and there is no such thing as a perpetual motion machine that produces energy eternally. In just the same way, area simply cannot be increased and decreased.

The dimensions of this puzzle were shown in Figure 4, but it's not necessary to construct the puzzle using these dimensions. The conditions of this magic formation are shown as parameters in Figure 7. If the length of the square's edge is x cm, then it is suitable for the rectangle to have a height of x cm, a width of a cm, and for the height of the right-angled triangle to be $2x$ cm. The vertical length of the playing card is not magically related. With these dimensions a parallelogram fissure with a base of $x^2/(2a+x)$ cm and a height of $2a+x$ cm is created. If $x \to 0$ in this diagram, the two trapezoids become congruent rectangles, and the whole thing fits together correctly.

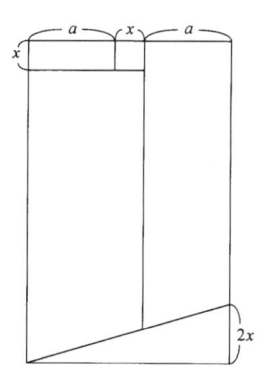

Figure 7: The Conditions for Magic

And finally, there is one more ingenious aspect to this piece of magic. It is in the shape of the card. Real playing cards have their corners cut into round curves. However, if the corners are cut then the trick will be given away when the card is flipped over with the left and right pieces exchanged. The card is shaped with square corners, but the design itself on the playing card expresses the curvature of the corners. I think the readers who have made it this far are potential magicians...

Reference

[1] I. Sakane, *Shin Asobino Hakubutushi* [*New Natural History of Puzzles*], Tokyo: Asahi Newspaper. (includes a description of [Sunflowers Facing the Sun] by H. Kondo), (1986).

CHAPTER **15**
The Mysterious Number 6174

Abstract: The number 6174 is a really mysterious number. At first glance, it might not seem so obvious, but as we are about to see, anyone who can subtract can uncover the mystery that makes 6174 so special.

AMS Subject Classification: 11A02, 00A09, 97A20
Key Words: 6174, Kaprekar operation, Kaprekar constant, Number theory

1 The Kaprekar Operation

6174 is truly a strange number. It is also a number with which we share a close relationship from elementary school up to university. But before explaining what kind of number it is, would you mind doing a little simple arithmetic?

First choose a single 4-digit number. When choosing, please avoid numbers with four identical digits like 1111 or 2222. For example, let's consider the year, 2005. Take the four digits that compose the number and reorder them into the largest and smallest numbers possible. For numbers with less than 4-digits, pad the left-hand side with zeroes to maintain 4-digits. In the case of 2005, the results are 5200 and 0025. Taking the difference between the largest and the smallest yields
$5200 - 0025 = 5175$.
This type of operation is known as a Kaprekar operation. The name derives from that of the Indian mathematician D.R. Kaprekar, who discovered a special property of the number 6174. Iterating the operation on our newly revealed number yields,

$7551 - 1557 = 5994$
$9954 - 4599 = 5355$
$5553 - 3555 = 1998$
$9981 - 1899 = 8082$
$8820 - 0288 = 8532$

$$8532 - 2358 = 6174$$
$$7641 - 1467 = 6174.$$

When the number becomes 6174, the operation repeats and 6174 thus cycles, and is known as the 'kernel.' No matter what the initial number may be, the sequence will eventually arrive at 6174. In fact, the kernel number 6174 will definitely be reached. If you remain doubtful, try the process again with a different number. 1789 develops as follows.
$$9871 - 1789 = 8082$$
$$8820 - 0288 = 8532$$
$$8532 - 2358 = 6174$$

2005 reaches 6174 after the Kaprekar operation is applied 7 times. 1789 reaches it after 3 times. This works for all 4 digit numbers. Isn't this strange? For elementary school pupils this is good practice for subtracting 4 digit numbers. For university students, thinking about why this happens reveals that 6174 is an exceptionally fascinating number. From this point on, I'd like to take a close look at the background of this number.

2 Solution Using Simultaneous Linear Equations

Let the largest number formed by rearranging the 4 digits be represented as *abcd* and the smallest as *dcba*. Since the solution is cyclic, the difference between these two may be expressed as a combination of the digits $\{a, b, c, d\}$.

With $9 \geq a \geq b \geq c \geq d \geq 0$ and the subtraction

$$
\begin{array}{r}
a\ \ b\ \ c\ \ d \\
-\ \ d\ \ c\ \ b\ \ a \\
\hline
A\ \ B\ \ C\ \ D
\end{array}
$$

the differences between each digit obey the following relationships.
$D = 10 + d - a\ (a > d)$
$C = 10 + c - 1 - b = 9 + c - b\ (b > c - 1)$
$B = b - 1 - c\ (b > c)$
$A = a - d$

Consider the relationship between the value of (A, B, C, D) and the set $\{a, b, c, d\}$. Since there are 4 equations and 4 variables, this is a 4-dimensional simultaneous linear equation. It ought to have a solution. Calculating the number of permutations of the elements of $\{a, b, c, d\}$ yields $4! = 24$ alternatives. It is sufficient to test each of these. The

details are omitted but the unique solution of this simultaneous linear equation occurs when $(A, B, C, D) = (b, d, a, c)$. Solving this we obtain $(a, b, c, d) = (7, 6, 4, 1)$.
$abcd - dcba = bdac$, i.e., $7641 - 1467 = 6174$, and the kernel number is 6174.

This phenomenon occurring with 4-digit numbers is also known to occur with 3-digit numbers. For example, with the 3-digit number 753, the calculation is as follows.

$753 - 357 = 396$
$963 - 369 = 594$
$954 - 459 = 495$
$954 - 459 = 495$

In the 3-digit case, the number 495 is reached, and this occurs for all 3-digit numbers. Why don't you try some?

3 The Number of Iterations Needed to Reach 6174

I first heard about the number 6174 from an acquaintance in around 1975 and it made a strong impression on me. I was surprised by the beautiful fact that all 4-digit numbers reach 6174, and thought it might be possible to prove this easily using high school-level mathematical knowledge. But the calculations are surprisingly complex and I left the problem in an unsolved state. At that time I made a copy of a journal paper about the topic. I attempted to investigate the upper limit on the number of iterations necessary to settle on the number 6174 using a computer. By means of a Visual Basic program with about 50 lines, I tested all the 4-digit numbers between 1000 and 9999. This included all 8991 digit natural numbers excluding those with 4 equal digits (1111, 2222, ⋯ ,9999).

Table 1 shows the frequency of each number of iterations needed to reach 6174. The largest number of steps needed is 7. If 6174 is not reached in 7 iterations, then a mistake was made during calculation. This is useful educational material for elementary school students to practice subtracting 4-digit numbers. For the initial number 6174, without even performing any Kaprekar operations, 6174 has already been reached, so in this case the number of iterations is taken as 0.

Number of iterations	Frequency
0	1
1	356
2	519
3	2124
4	1124
5	1379
6	1508
7	1980
Total	8991

Table 1. Number of Iterations Needed to Reach 6174

4 The Route to 6174

Kaprekar was an Indian mathematician active in the 1940s, and you can find more details about him in [2]. The aspects of the problem are explained as follows.

Taking an arbitrary 4-digit number expressed as *abcd* (where $a \geq b \geq c \geq d$) and executing the first subtraction may be considered as follows. The largest 4-digit number is equal to $1000a + 100b + 10c + d$, so the smallest number is $1000d + 100c + 10b + a$. Subtracting the smallest number from the largest and combining similar terms yields the following.

$1000a + 100b + 10c + d - (1000d + 100c + 10b + a)$
$= 1000(a - d) + 100(b - c) + 10(c - d) + (d - a)$
$= 999(a - d) + 90(b - c)$

Here, $a - d$ has a value between 1 and 9, and $b - c$ takes an arbitrary value between 0 and 9, so in total there are 90 numbers taking the form above. Figure 1 was produced in order to confirm this fact.

In this figure $(a - d) \geq (b - c)$, so the 36 entries in the bottom left (a catch-all case) are meaningless numbers. Next we will execute the second subtraction, so the numbers in Figure 1 are rearranged into the corresponding largest values as shown in Figure 2.

Ignoring the repetitions of the catch-all cases, there are 30 entries remaining in this figure. Figure 3 shows a schematic diagram of the ways in which these 30 numbers reach 6174. According to this diagram, it should be possible to understand at a glance that all 4-digit natural numbers reach 6174. It can also be seen that at most 7 iterations are necessary. Even so, it's certainly strange. Was Kaprekar, who discovered

		|	999X(a-d)							
		| 1	2	3	4	5	6	7	8	9
	0	999	1998	2997	3996	4995	5994	6993	7992	8991
	1	1089	2088	3087	4086	5085	6084	7083	8082	9081
	2	1179	2178	3177	4176	5175	6174	7173	8172	9171
	3	1269	2268	3267	4266	5265	6264	7263	8262	9261
90X	4	1359	2358	3357	4356	5355	6354	7353	8352	9351
(b-c)	5	1449	2448	3447	4446	5445	6444	7443	8442	9441
	6	1539	2538	3537	4536	5535	6534	7533	8532	9531
	7	1629	2628	3627	4626	5625	6624	7623	8622	9621
	8	1719	2718	3717	4716	5715	6714	7713	8712	9711
	9	1809	2808	3807	4806	5805	6804	7803	8802	9801

Figure 1: The Numbers after the First Subtraction

		|	999X(a-d)							
		| 1	2	3	4	5	6	7	8	9
	0	9990	9981	9972	9963	9954	9954	9963	9972	9981
	1	9810	8820	8730	8640	8550	8640	8730	8820	9810
	2		8721	7731	7641	7551	7641	7731	8721	9711
	3			7632	6642	6552	6642	7632	8622	9621
90X	4				6543	5553	6543	7533	8532	9531
(b-c)	5					5544	6444	7443	8442	9441
	6						6543	7533	8532	9531
	7							7632	8622	9621
	8								8712	9711
	9									9801

Figure 2: The Numbers Prior to the Second Subtraction

this, a man of exceptional intelligence or was he a man of exceptional leisure?

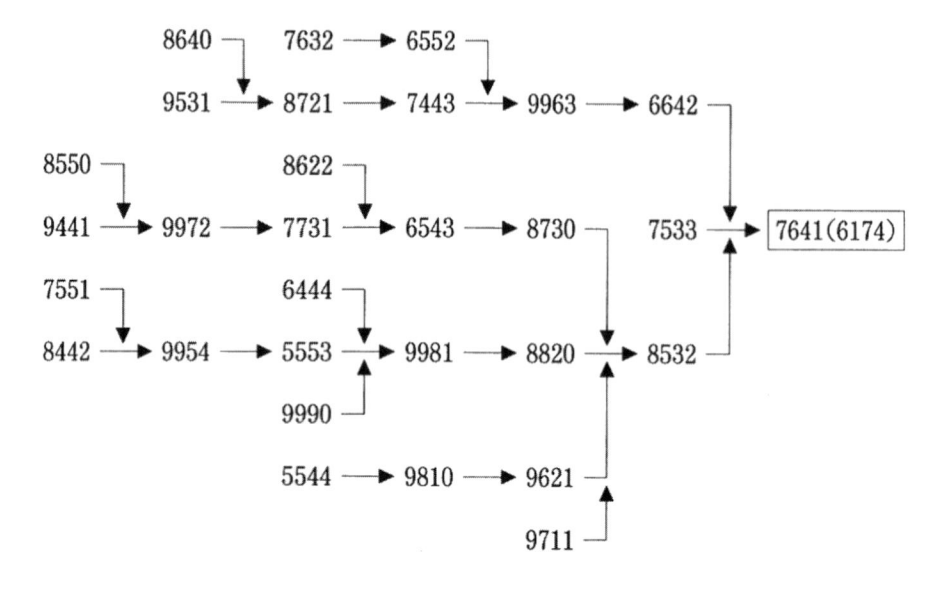

```
8640 ┐      7632 ──▶ 6552 ┐
     ▼                    ▼
     9531 ──▶ 8721 ──▶ 7443 ──▶ 9963 ──▶ 6642 ┐
                                               ▼
8550 ┐      8622 ┐                             
     ▼           ▼                             
9441 ──▶ 9972 ──▶ 7731 ──▶ 6543 ──▶ 8730 ┐  7533 ──▶ │7641 (6174)│
                                          ▼        ▲
7551 ┐      6444 ┐                        │        │
     ▼           ▼                        │        │
8442 ──▶ 9954 ──▶ 5553 ──▶ 9981 ──▶ 8820 ──▶ 8532 ┘
                 ▲
         9990 ──┘

         5544 ──▶ 9810 ──▶ 9621 ┐
                               ▲
                       9711 ──┘
```

Figure 3: Schematic Diagram Leading to 7641 (6174)

5 Recurring Decimals

Real numbers include both rational and irrational numbers. Rational numbers are those that can be expressed as a fraction $\frac{n}{m}$ (m, n are integers and $m \neq 0$), while irrational numbers are those that cannot be expressed in this way.

Real numbers may be written as decimals. Rational numbers are finite or recurring decimals. Irrational numbers, on the other hand, are non-cycling infinite decimals. For example, the following numbers are irrational.

$\sqrt{2} = 1.41421356237309504880168872420\cdots$
$\pi = 3.14159265358979323846264338327\cdots$

Recurring decimals, which are rational numbers, are explained as follows. Recurring decimals are those infinite decimals for which after some point a sequence of digits (the recurring sequence) repeats indefinitely. When writing recurring decimals, a dot is marked above each end of the recurring sequence when it first appears. The remaining digits are omitted. For example,

$0.7\dot{2}1\dot{4} = 0.7214214214\cdots$
is such a recurring decimal. By expressing this number using a geometrical progression, and using the formula for geometrical progression, it is possible to find a corresponding fraction.

$$0.7\dot{2}1\dot{4} = \frac{7}{10} + \frac{214}{10^4} + \frac{214}{10^7} + \cdots$$

$$= \frac{7}{10} + \frac{214}{10^4}(1 + \frac{1}{10^3} + \frac{1}{10^6} + \cdots)$$

$$= \frac{7}{10} + \frac{214}{10^4} \times \frac{1}{1 - \frac{1}{10^3}}$$

$$= \frac{7}{10} + \frac{214}{10(10^3 - 1)} = \frac{7209}{9990}$$

The denominator is $9990 = 2 \times 3^3 \times 5 \times 37$.

This recurring decimal is a rational number whose denominator contains other prime factors besides 2 and 5. Rational numbers may be classified as follows by examining the prime factorization of the denominator.

Finite decimals: factors include 2 and 5 alone
Pure recurring decimals: factors do not include 2 or 5
Mixed recurring decimals: factors include 2 or 5 as well as other factors

Pure recurring decimals are formed from the recurring sequence alone, while mixed recurring decimals also include another part besides the recurring sequence. For example, $\frac{1}{4} = 0.25$ is a finite decimal, $\frac{1}{7} = 0.\dot{1}4285\dot{7}$ is a pure recurring decimal and $\frac{1}{12} = 0.08\dot{3}$ is a mixed recurring decimal, because $4 = 2^2, 7 = 7$ and $12 = 2^2 \times 3$.

The Kaprekar operation may be applied to these recurring decimals. When the number has 3 or 4 digits, after a finite number of iterations the numbers 495 and 6174 are reached, respectively, so the sequence has a form like a single mixed recurring decimal.

6 What Happens with 2 Digits and with 5 or More Digits?

Numbers with 4 or 3 digits converge on a unique number, but what happens in the case of 2 digits? For example, starting with 28 and repeating the largest-minus-smallest Kaprekar operation yields

$28 \to 82-28=54 \to 54-45=9 \to 90-09=81 \to 81-18=63 \to 63-36=27 \to 72-27=45 \to 54-45=9,$

which beginning with 9, cycles in the pattern $9 \to 81 \to 63 \to 27 \to 45 \to 9$. Thus for numbers of 2 digits, a certain domain cycles in a similar fashion to two mixed recurring decimals.

Next, what happens with 5-digit numbers? First, isn't there a kernel value like 6174 and 495? Expressing a 5-digit number using $9 \geq a \geq b \geq c \geq d \geq e \geq 0$, the largest minus the smallest may be written as $abcde - edcba = ABCDE$, where (A, B, C, D, E) is chosen from the 120 permutations of $\{a, b, c, d, e\}$. This is a constrained case-based simultaneous linear equation. Regarding the 5-digit Kaprekar problem, a considerable amount of computation has already been performed and as a consequence it is known that there is no kernel value, and all 5-digit numbers enter one of the following loops.

$71973 \to 83952 \to 74943 \to 62964$
$75933 \to 63954 \to 61974 \to 82962$
$59994 \to 53955$

Regarding integers with 6 or more digits, Malcolm Lines indicates that increasing the number of digits soon becomes a tedious issue merely increasing the computation time [2]. The existence of kernel values is summarized in Table 2.

Digits	Kernel values	
2	Nothing	
3	495	Unique
4	6174	Unique
5	Nothing	
6	549945, 631764	
7	Nothing	
8	63317664, 97508421	

Table 2. Kernel Values

This table reveals that for 6 and 8 digits, there are 2 kernel values, and in some cases one of the kernel values is reached, while in others cases the sequence enters a loop. For a computer with 32-bit words, integers are represented using 32 bits, so it is possible perform calculations up to $2^{31} - 1 (= 2147483647)$, which is around the beginning of 10-digit numbers. Just as Lines described, it began to seem nonsensical, so I stopped calculating.

I wanted to know more about the roots of this problem and investigated a little further. I encountered Martin Gardner's book by chance, and came to understand the situation [1]. The explanation at the end of his book states that the number 6174 is called Kaprekar's constant after an Indian named Dattathreya Ramachandra Kaprekar, that he was the first person to demonstrate its importance ("Another Solitaire Game", *Scripta Mathematica*, 15 (1949), pp 244-245), and that he subsequently presented "An Interesting Property of the Number 6174"(1955), "The New Constant 6174"(1959) and "The Mathematics of the New Self Numbers"(1963). It seems that these represent the truth of the matter, and computer-based revelations have placed the number under a spotlight of attention.

Martin Gardner takes it that for numbers with 1, 2, 5, 6 and 7 digits, Kaprekar's constant does not exist. However, as shown above, for 6-digit numbers there are two kernel values (these cases are not known as Kaprekar's constants). Gardner also states kernel values for numbers with 8, 9 and 10 digits. For 8 digits it is 97508421, for 9 digits 864197532 and for 10 digits 9753086421. The subtractions are as follows.

98754210 − 01245789 = 97508421

987654321 − 123456789 = 864197532

9876543210 − 0123456789 = 9753086421

The form is beautiful in the 9- and 10-digit cases, and these were probably obtained intuitively. Perhaps the value for the 8-digit case was obtained through computer processing. It might alternatively have been found using a calculator or pencil and paper arithmetic. It is sufficient if there is a way to find out without resorting to computer power. If I have another opportunity I'll try investigating using a different method. This problem remains captivating, and I'd like to mention David Wells' book which is of historical interest and discusses each of the numbers [4].

7 Chance or Necessity?

The existence of cyclic numbers is made clear by the simultaneous linear equations. It was proven that for 3-digit numbers 495 is a unique cyclic number, and likewise 6174 for 4-digit numbers. It was also confirmed that all 3-digit numbers converge on 495 and that all 4-digit numbers converge on 6174. However, this was merely demonstrated, and I think that the real reason why all the numbers converge on the cyclic number has not been demonstrated.

Is it by chance or by necessity that it only works for 3- and 4-digit numbers? I have a feeling that it is a matter of chance. Allow me to introduce the following puzzle which has already been solved [3].

$$\begin{array}{r} \square\square\square\square\square \\ \times\ \square\square\square\square\square \\ \hline 1\ 2\ 3\ 4\ 5\ 6\ 7\ 8\ 9 \end{array}$$

This worm-eaten arithmetic puzzle involves putting numbers in the blank boxes, but the form is so beautiful that I had hoped in my heart that perhaps some great theory of numbers lay hidden within, but I found out that it is merely coincidental. It has been confirmed that there is a host of such puzzles in existence.

$$\begin{array}{r} \square\square\square\square\square \\ \times\ \square\square\square\square\square \\ \hline 1\ 2\ 3\ 4\ 5\ 6\ 7\ 8\ 4 \end{array}$$

Kaprekar's problem may be thought of as a similar type to this worm-eaten arithmetic problem. The trick for solving this problem is to use the prime factorization. Applying this to the first example,

$123456789 = 3 \times 3 \times 3607 \times 3803$,

and the answer is 10821×11409. Applying the same method to the latter,

$123456784 = 2^4 \times 11^2 \times 43 \times 1483$,

and the answer is 10406×11864.

Historically, some of the developments in science and mathematics have been prompted by 'mistakes.' For the worm-eaten arithmetic problem, given the former case the desire to try and solve it occurs naturally, while in the latter case one probably wouldn't be particularly interested. The reason is that the former statement appears so beautiful. Just knowing that under Kaprekar operations all 4-digit numbers converge on 6174 and all 3-digit numbers on 495 provides sufficient charm as a mathematical puzzle. Who could say that this is merely a coincidence? I hope that some great theory of mathematics might lie behind it. This hope might end up being a beautiful mistake, but that's something I don't wish to believe.

References

[1] M. Gardner, (trans. By S. Hitotsumatsu), *Matrix Hakaseno Kyoino Suhijyutsu* [*The Magic Numbers of Doctor Matrix*], Tokyo: Kinokuniya, (1978).

[2] M. Lines, (trans. By K. Katayama), *Su: Sonoigaina Hyojyo* [*A number for your thoughts ? Facts and speculations about numbers from Euclid to the latest computers*], Tokyo: Iwanami, (1988).

[3] Y. Nishiyama, Utsukushii Puzzles [Beautiful Puzzles], *Kurashino Algorizumu* [*Algorithms of Living*], Kyoto: Nakanishiya, (1989).

[4] D. Wells, (trans. By N. Yoshigahara), *Su no Jiten* [*The Penguin dictionary of Curious and Interesting Numbers*], Tokyo: Tokyo Books, (1987).

CHAPTER 16
Numerical Palindromes and the 196 Problem

Abstract: The numbers 727, 1991, 38483 and so on are the same when read forwards or backwards. These numbers are known as numerical palindromes. Let's pick an arbitrary number, reorder the digits in reverse and add it to the original number. It is said that repeating this operation eventually leads to a palindrome. Numbers in the sequence derived from 196, however, do not yield a palindrome. This is an unresolved mathematical problem.

AMS Subject Classification: 11A02, 00A09, 97A20
Key Words: Numerical palindromes, 196 problem, 196 and 879, Seed

1. Numerical Palindromes

There is a certain concept referred to as a 'palindrome.' Palindromes are phrases that are identical when read both forwards and backwards. In Japanese, the phrases "TA KE YA BU YA KE TA" and "U TSU I KE N SHI HA SHI N KE I TSU U" are palindromes (meaning "The bamboo bush burned" and "Mr. Ken Utsui has neuralgia," respectively). There are also some outstanding literary creations such as "NA KA KI YO NO TO O NO NE FU RI NO MI NA ME SA ME NA MI NO RI FU NE NO O TO NO YO KI KA NA" which takes the form of a traditional 31-syllable Japanese poem known as a tanka. It is written "長き夜の遠の眠りの皆目覚め波乗り船の音の良きかな" and means "For those who awaken after a long night of deep sleep, how comforting is the sound of a boat on the waves."

Numbers can also be palindromes. For example, 727, 1991, 38483 and so on are the same when read forwards or backwards. Numbers which are symmetrical in this way are known as numerical palindromes.

Now, let's pick an arbitrary number, reorder the digits in reverse and add it to the original number. It is said that repeating this operation

eventually leads, at some point, to a palindrome. For example, suppose we pick 59.
$$59 + 95 = 154, \quad 154 + 451 = 605, \quad 605 + 506 = 1111$$
Let's try again with another number, 183.
$$183 + 381 = 564, \quad 564 + 465 = 1029, \quad 1029 + 9201 = 10230,$$
$$10230 + 3201 = 13431$$
After 3 repetitions of the operation, the number 59 arrived at the palindrome 1111. After 4 repetitions, the number 183 arrived at the palindrome 13431. Please try this for yourself with another number.

Almost all numbers arrive at a palindrome when this operation is repeated, but numbers in the sequence derived from 196 do not arrive at a palindrome. Also, for a given number, it is not known whether it will arrive at a palindrome. This problem is both old and new. It has been discussed in *Scientific American*, and although I wasn't particularly interested at that time, I later developed an interest which prompted me to investigate it in this chapter.

2. All 2-digit Numbers Lead to Palindromes

As a starter, let's try some 2-digit numbers. It can be confirmed that starting from any 2-digit number from 10 to 99 leads to a palindrome. The number 89 doesn't seem to yield a palindrome, but eventually after 24 iterations, the following 13-digit number is the first palindrome obtained.

8813200023188

Meticulously investigating every one of the 2-digit numbers to confirm that it yields a palindrome is not very mathematical. Paying attention to certain digits and observing their combined sum reveals that the result is a palindrome if all the sums are less than or equal to 9. For example, for 35, $3 + 5 = 8$, which is less than 9 so it is not necessary to investigate further. There are 90 2-digit numbers, from 10 up to 99. These numbers can be written $ab(1 \leq a \leq 9, 0 \leq b \leq 9)$.

(1) When the units' digit is 0, the result is a palindrome. There are 9 such cases. ($a0$)

(2) When the units' digit and the tens' digits are the same, the number is already a palindrome. There are 9 such cases. (aa)

(3) When the units' digit and the tens' digit are symmetrical there is no need to investigate any further. There are 36 such cases. (ab, ba)

(4) When the sum of the units' digit and the tens' digit is less than or equal to 9 there is no carry to another digit, so the result is a palindrome.

There are 16 such cases. ($a + b \leq 9$)

(5) When the sum of the units' digit and the tens' digit is less than or equal to 13, the result is a 3-digit number and may be denoted abc. Then each digit is less than or equal to 4, so in the next step it is certain to yield a palindrome. There are 14 such cases. ($10 \leq a + b \leq 13$)

(6) When the units' digit and the tens' digit sum to 14, the numbers may be 59 or 68. Both $59 + 95 = 154$, and $68 + 86 = 154$ so it is sufficient to investigate only one case. Likewise, when the sum is 15, the number may be 69 or 78, but $69 + 96 = 165$ and $78 + 87 = 165$, so again it is sufficient to investigate only one case.

According to the results above, since $90 - 9 - 9 - 36 - 16 - 14 - 2 = 4$, it is sufficient to investigate only the following 4 cases.

$$59, 69, 79, 89$$

Drawing up a table of these observations reveals the following (Figure 1). The 4 cases mentioned above are located in the lower right-hand side of the table. The numbers that are difficult to turn into palindromes are those for which the units' digit and the tens' digit are close to 9. It is therefore predictable that it is difficult with the number 89.

10	11	12	13	14	15	16	17	18	19
20	21	22	23	24	25	26	27	28	29
30	31	32	33	34	35	36	37	38	39
40	41	42	43	44	45	46	47	48	49
50	51	52	53	54	55	56	57	58	59
60	61	62	63	64	65	66	67	68	69
70	71	72	73	74	75	76	77	78	79
80	81	82	83	84	85	86	87	88	89
90	91	92	93	94	95	96	97	98	99

Figure 1: Validating the 2-Digit Numbers

3. The 3-Digit Numbers 196 and 879

Next, let's consider 3-digit numbers. A similar method to that used for 2-digit numbers can be used for 3-digit numbers, but along with the increase in the number of digits, the degree of refinement gets worse. There are 900 numbers with 3-digits, from 100 to 999. At present the following 13 from among them are known to not yield palindromes.

$$196, 295, 394, 493, 592, 689, 691, 788, 790, 879, 887, 978, 986$$

The first of these numbers which do not yield palindromes is 196, so this issue is also known as the '196 problem.' I found this an interesting problem and began to investigate myself, but I needed the help of a computer program. Table 1 shows the number of iterations needed to convert the numbers into a palindrome.

Among the 900 cases, 90 of them are already palindromes so an investigation is not needed. Investigating the process by which the remaining 810 numbers yield palindromes, there are 213 which yield palindromes in 1 step, 281 in 2 steps, and 145 which require 3 steps. This reveals that there are many numbers which yield palindromes in an unexpectedly small number of steps. The slowest require 23 steps, and there are 7 such cases. There are also the 13 cases mentioned above which do not yield palindromes.

The 13 3-digit numbers that do not yield palindromes can be divided into two groups. The first group contains 196, 295, 394, 493, 592, 689, 691, 788, 790, 887 and 986, and the second group contains 879 and 978. These are shown in the schematic diagram in Figure 2. The numbers 196 and 879 are known as 'seeds,' and it can be seen that all the other numbers can be represented by these seeds. The reason is because the number 691 is the reverse ordering of the number 196 so they are members of the same group ($196 + 691 = 691 + 196 = 887$). 295 and 592 yield the same number, 887, after one iteration so they are also in the same group ($295 + 592 = 196 + 691 = 887$).

I do not know whether these two groups are forever separate, or whether they eventually unify.

Iterations	Frequency	Proportion
0	90	10.0%
1	213	23.7%
2	281	31.2%
3	145	16.1%
4	63	7.0%
5	31	3.4%
6	9	1.0%
7	17	1.9%
8	7	0.8%
10	2	0.2%
11	7	0.8%
14	2	0.2%
15	7	0.8%
17	4	0.4%
22	2	0.2%
23	7	0.8%
Over 100	13	1.4%
Total	900	100%

Table 1. The Number of Iterations Required to Reach a Palindrome (3-Digit Numbers)

4. The Root of the 196 Problem

I wonder when the number 196 was first recognized as problematic with regard to numerical palindromes. Regarding the literature in Japan, 'Asimov's Collection of General Knowledge' [2] contained an explanation about 196. The original document corresponding to the Japanese translation was published in 1979, so it is clear that the topic has been known about for quite some time [1]. Gardner and other authors also took up the topic a number of times in *Scientific American* around this time.

I was wondering whether it was Asimov who first mentioned it, and therefore investigated a little further. I wrote this manuscript while I was visiting Cambridge, so I was able to lay my hands on some precious documents in the library there. They revealed that Trigg had already taken up the problem of 196 in *Mathematics Magazine* in 1967 [6], and that also Lehmer had raised the issue in *Sphinx* magazine published in Brussels in 1938 [5]. He had performed 73 iterations without reaching a

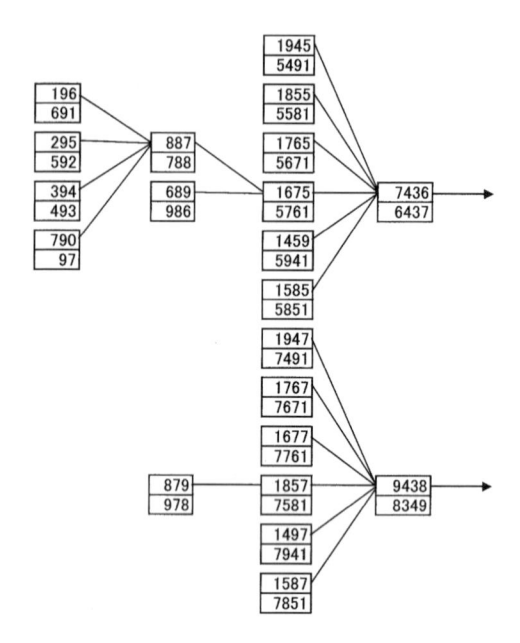

Figure 2: Schematic Diagram Showing the 13 Unresolved Cases for 3-Digit Numbers

palindrome. It may be possible to trace back even further, but I could find no other existing documents.

The Japanese word for palindrome is *kaibun*. The word 'palindrome' originated in the 17th century, and it was adopted from Greek. The period when Descartes and Newton were active was between the 17th and 18th centuries, and although it cannot be confirmed in the literature, it is possible that the problem of 196 was a topic of discussion among the mathematicians of this time. At any rate, in the 70 years or so since 1938, many mathematicians have pitted themselves against the problem of 196, and it is as yet a long-lived unsolved problem.

5. A World Record and Still Counting

Let's take a look at the records achieved by the mathematicians who have wrestled with this problem. In 1938 Lehmer calculated 73 iterations starting with 196 without reaching a palindrome, and obtained the following 35 digit number.

45747 6603920132 8565933091 8416673654

This was the highest record at that time. When I repeated the calculation using a Visual Basic program, I found the following slightly different number.

45747 6591819132 8565933092 7106673654

Speaking of 1938, it was a time when computers had not yet appeared, and printed on the back cover of the magazine there was an advertisement for a desktop calculator, *i.e.*, a cash register. It could handle up to 12 digits. The mathematicians of that time pitted themselves against the problem of 196 using tools that could only handle calculations up to 12 digits! In 1967 Trigg computed 3556 iterations yielding a 1700 digit number and confirmed that no palindrome was reached. He used the latest computer of the time, which was an IBM1401.

More recent data was provided by Walker in 1990, who computed 2,415,836 iterations, obtained a 1,000,000-digit number and confirmed the absence of a palindrome. At present 1,000,000 has been exceeded, and since the numbers are so long I will express them in terms of millions. In February 2006 Landingham completed 699 million iterations achieving a 289-million digit number without finding a palindrome, and the computation continues [4]. In order to show just how big a number this is, expressing it using an exponent reveals that although it is as large as $10^{289,430,478}$, there is still no palindrome. Since $289 \div 699 = 0.413\cdots$, each calculation increases the number of digits by approximately 0.4, or roughly 1 digit for every 2 iterations.

The original version of Asimov's book was published in 1979, which was before a million iterations had been achieved. It is now almost 30 years since then. Technical innovation in computing has advanced, and the capabilities of home computers have improved, but despite reaching 289-million digits the problem remains unsolved.

The discussion above concerns the world record for which the number 196 has been confirmed as not yielding a palindrome, but there is another record. It is shown in Table 2, and is the largest number of iterations required by a number before it eventually *does* yield a palindrome.

This table is read as indicating that the 2-digit number 89 requires 24 iterations before it yields a palindrome, and that the 3-digit number 187 requires 23 iterations. Among the largest numbers of iterations so far obtained, the current world record is for the 17-digit number 10,442,000,392,399,960 which requires 236 iterations. The record is held by Doucette and was calculated in 2005 [3].

Digits	Number	Number of iterations
2	89	24
3	187	23
4	1,297	21
5	10,911	55
6	150,296	64
7	9,008,299	96
8	10,309,988	95
9	140,669,390	98
10	1,005,499,526	109
11	10,087,799,570	149
12	100,001,987,765	143
13	1,600,005,969,190	188
14	14,104,229,999,995	182
15	100,120,849,299,260	201
16	1,030,020,097,997,900	197
17	10,442,000,392,399,960	236

Table 2. The Largest Number of Iterations Required to Yield a Palindrome [3]

6. Will the Problem of 196 Ever Be Solved?

The world record is still open, but will the problem of 196 eventually be solved? At this point let's shift our perspective and consider what happens to the ratio of numbers that are palindromes when the number of digits increases. There are 9 1-digit numbers from 1 to 9, and these can all be regarded as palindromes. There are 90 2-digit numbers from 10 to 99, and there are 9 palindromes, 11, 22, 33, 44, 55, 66, 77, 88 and 99. The ratio of palindromes is $9/90 = 0.1$.

There are 900 3-digit numbers between 100 and 999, and the number of palindromes can be calculated as follows. The 3-digit palindromes all take one of the following forms.

$$1x1, 2x2, 3x3, 4x4, 5x5, 6x6, 7x7, 8x8, 9x9$$

There are 10 digits from 0 to 9 that can be substituted for x, so the number of palindromes is $9 \times 10 = 90$. The ratio of palindromes is $90/900 = 0.1$.

For 4-digit numbers, there are 9000 between 1000 and 9999. Palindromes with 4-digits have one of the following forms.

$1xx1, 2xx2, 3xx3, 4xx4, 5xx5, 6xx6, 7xx7, 8xx8, 9xx9$

There are 10 ways of replacing with the digits 00 to 99, so the number of palindromes is $9 \times 10 = 90$. The ratio of palindromes is thus $90/9000 = 0.01$. The ratios can be computed in the same way for 5 and 6 digits, and the results are shown in Table 3.

Digits	Total	Palindromes	Ratio
2	90	9	0.1
3	900	90	0.1
4	9000	90	0.01
5	90000	900	0.01
6	900000	900	0.001

Table 3. The Ratio of Palindromes

In general, for $2n$ digits the ratio of palindromes is $\frac{1}{10^n}$. When the number of digits is $2n+1$, the ratio is the same as for $2n$. The ratio of palindromes can therefore be summarized as $\frac{1}{10^n}$ for both $2n$-digit (an even number of digits) and $2n+1$-digit (an odd number of digits) numbers.

Thus, as the number of digits increases, for every two new digits the proportion of palindromes decreases by a factor of 10. The number 196 has at present been taken as far as 289 million digits, and I suppose you can imagine just how small this makes the ratio of palindromes. However, no matter how small the ratio, palindromes do exist, and that's why it's such a frustrating problem for mathematicians.

The ratio of palindromes certainly grows extremely small, but before this happens there are many addition operations. There are 81 2-digit numbers that are not palindromes, but among them 49 (60%) yield a palindrome after one operation. These are good odds. For the 3-digit numbers, 810 of them are not palindromes, and 213 (26%) of these yield a palindrome after one operation. The proportion that yield a palindrome after one operation surely decreases as the number of digits increases, but compared to the ratio of palindromes discussed above it is large.

At present the calculations for 196 following on from 289 million digits are continuing on and on, but whether those calculations are drawing closer to a palindrome or whether they will continue forever as non-palindromes is unclear. Neither is it known through what states the numbers transition. The four-color problem was ultimately solved by resorting to the power of computers, but wouldn't it be nice if there were a more mathematical approach that did not require their use? I certainly hope those readers whose interest has been piqued by numerical palindromes and the problem of 196 will attempt the challenge of finding a proof.

References

[1] I. Asimov, *Isaac Asimov's Book of Facts*, New York: Grosset and Dunlap, (1979).

[2] I. Asimov, (trans. S. Hoshi) *Asimov no Zatsugaku Collection [Asimov's Collection of General Knowledge]*, Tokyo: Shinchosha, (1986).

[3] J. Doucette, *196 Palindrome Quest, Most Delayed Palindromic Number*, (2005).

http://www.jasondoucette.com/worldrecords.html.

[4] W.V. Landingham, *196 and Other Lychrel Numbers*, (2006).

http://www.p196.org/.

[5] D. Lehmer, Sujets d' étude, *Sphinx* (Bruxelles), 8(1938), 12-13, (1938).

[6] C.W. Trigg, Palindromes by Addition, *Mathematics Magazine*, 40(1967), 26-28, (1967).

CHAPTER 17
From Oldham's Coupling to Air Conditioners

Abstract: Oldham's coupling is a really interesting device. It is possible to understand the method with a knowledge of only junior high-school geometry. Compressors in air conditioners also use this idea, where involute curves are used for the teeth of the cogs in the compressor. Mathematics doesn't just reside in textbooks; it's alive in our daily lives.

AMS Subject Classification: 51M05, 00A09, 97A20
Key Words: Oldham's coupling, Turning block double-slider mechanism, Air conditioner compressors, Involute curves in cogs

1. In a Certain Museum...

Mathematics is not just to be found in textbooks, it exists in our daily lives. It does not just have abstract forms, but is also tangible and constant. Once when I was visiting Kyoto University Museum and wondering whether there were any interesting educational materials, I came across a model of a device known as Oldham's coupling and it caught my attention. Oldham's coupling is a design that was imported into Japan through Germany at the time of modernization in the 19th and 20th centuries. Oldham is apparently the name of the man who devised it. I was able to touch the object, and I was captivated by its peculiar and wondrous motion.

It has two parallel axes, which are slightly offset. How can the rotation of the left axis be correctly transmitted to the rotation of right axis? The idea of using three cogs may occur to an amateur. There would be two cogs with the same number of teeth, and one for changing the direction of motion. A solution is thus possible with a total of three cogs. However, when the axial distance is very small, correspondingly tiny cogs are required, which is not realistic. There is the so-called 'universal joint' used in cars, and while it is possible to solve the problem in this way, the device ends up being rather complicated. It is also possible to apply a belt, but belts stretch, shrink and wear down, so the rotation is not transmitted correctly. 'Oldham's coupling' which I will introduce here, is extremely mathematical, but it does not require high-level mathematics. Rather, it is possible to understand the method with a knowledge of only junior high-school geometry.

2. Transmitting Rotation between Parallel Axes

The structure of Oldham's coupling is shown in Figure 1, which is a reproduction based on the work of Hitoshi Morita [1].

It is composed of 3 discs, a, b and c. Applying a rotation to disc a or c causes disc b to rotate while sliding with respect to a and c. This mechanism is known as a turning block double-slider crank mechanism, and b is called a double-sliding crank. The discs a and c have grooves cut along their diameters, and as shown in Figure 1(b), b has two corresponding projections which are at right angles to each other, and fit into the grooves in both a and c. When a rotates through a given angle, b and c also rotate through the same angle so the angular velocities of a and c are equal.

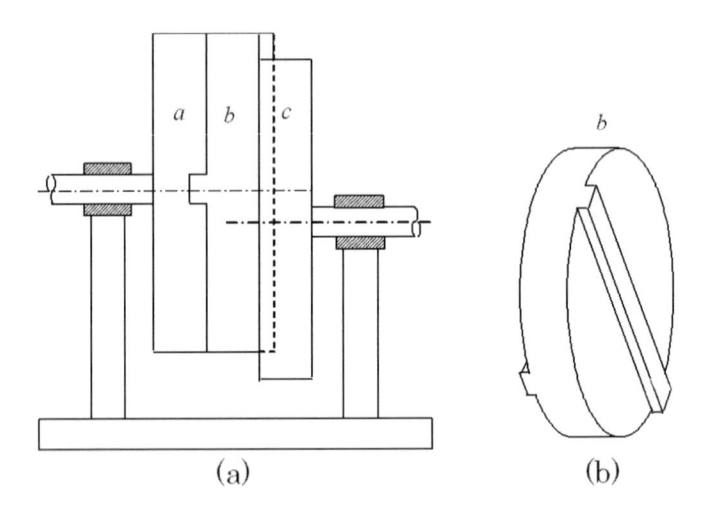

Figure 1: Turning Block Double-slider Mechanism

The discs a and c on the left and right sides of Oldham's coupling rotate in circles at the same velocity, but the rotation of disc b is neither circular nor ellipsoidal, and its motion is peculiar. It moves according to intermediate parameters, like a cycloid or trochoid. Let's confirm this mathematically below.

The 3 discs are modeled as shown in Figure 2. The point on the end of the groove in disc a is denoted P, and the point at the end of the groove in disc c is denoted Q. P rotates with a uniform circular motion about center O_1 with radius r. Q rotates with a uniform circular motion with radius r about O_2, which is only separated from O_1 by a distance d. The coordinates of $P(x, y)$ and $Q(x, y)$ are as follows.

$$P(r\cos(\frac{\pi}{2} - \theta), r\sin(\frac{\pi}{2} - \theta))$$

$$Q(r\cos(-\theta) + d, r\sin(-\theta))$$

The coordinates of the intersection point between the two perpendicular grooves is denoted $R(x, y)$, and this is the center, O_3, of disc b. Calculation yields

$$R(\frac{d}{2}(1 - \cos 2\theta), \frac{d}{2}\sin 2\theta).$$

Point R moves in a uniform circular motion with center $(\frac{d}{2}, 0)$ and radius $\frac{d}{2}$. Also, note that the angular velocity of this point R is twice that of the discs a and c.

The point $S(x, y)$ on the circumference of disc b is therefore as follows.

$$S(x, y) = R(x, y) + P(x, y)$$

Writing the angular velocity as ω, the time as t and the angle as θ, we have $\theta = \omega t$, and $P(x, y)$ moves with angular velocity ω. The center $R(x, y)$ moves with angular velocity 2ω, so it is clear that $S(x, y)$ does not have a constant angular velocity. It is possible to take the derivative in the x direction and in the y direction with respect to t, but it is not possible to express them with a simple equation in the same way as the cycloid (an equation like $f(x, y) = 0$). They cannot be expressed clearly using formulae, but taking a small value of δt, and numerically calculating the rate of change of $S(x, y)$, denoted $S'(\frac{dx}{dt}, \frac{dy}{dt})$, reveals that the motion does not have uniform angular velocity.

Furthermore, the orthogonality of the grooves is not an essential condition. It is only necessary for there to be an angle between them, and if this is the case then rotational motion can be correctly transmitted between the two axes. We won't delve deeply into this point here.

(a) The left and right discs

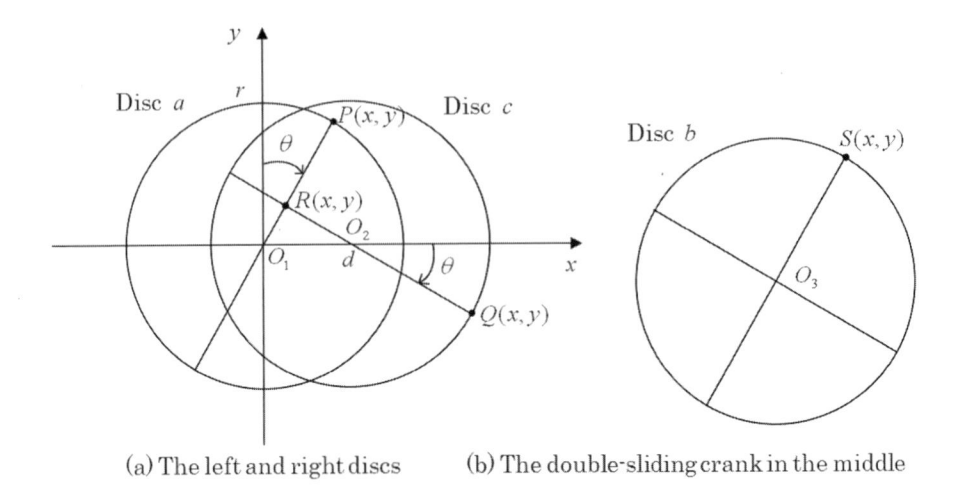

(b) The double-sliding crank in the middle

Figure 2: Coordinate System for Oldham's Coupling

The way the 3 discs move when θ changes from 0 to $\frac{\pi}{2}$ is summarized in Figure 3. In order to make the nature of the change clear, a point on the discs has been marked with a circle. The discs a and b are shown with the point at the edge of the groove, and disc c is only offset by $\frac{\pi}{2}$. The discs on the left and right, a and c, have their centers fixed so they move with simple circular motions. The center of the disc in the middle, b, is always moving, so its motion is complicated. The trajectory is neither a circular nor an elliptical orbit. Also, the speed is not constant. Initially, it begins from the position of disc a ($\theta = 0$), and advances while its velocity gradually increases. Finally it overlaps the position of disc c ($\theta = \frac{\pi}{2}$). The center of the curve enclosing the locus of motion for disc b is ($\frac{d}{2}, 0$), and it is a circle with radius $r + \frac{d}{2}$. Disc b therefore never protrudes outside this circle.

Having expressed the coordinates mathematically, it is possible to derive the formula and find the change of speed, but the equation is awkward so I have omitted it here. Even so, the complicated motion of disc b is surprising.

We were able to confirm the motion of Oldham's coupling mathematically, but does it really move so nicely? I started to want to make a model. First I made a model using thick drawing paper, but it didn't move very well. I therefore decided to make a model out of wood. The

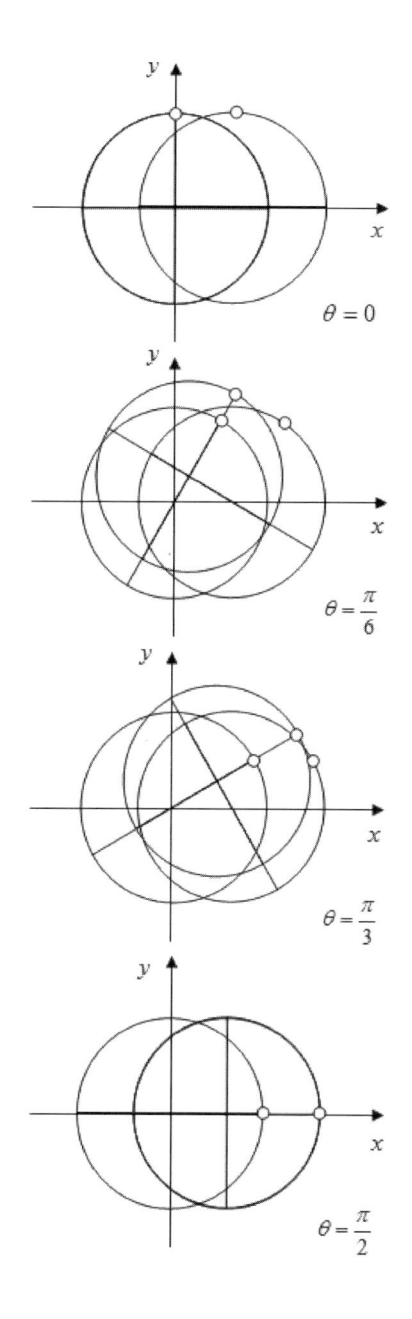

Figure 3: Transition Diagram ($0 \leq \theta \leq \frac{\pi}{2}$)

materials needed for construction were on sale in certain DIY shops, and the model I created and assembled is shown in Figure 4. The cost of materials was about 2000 yen. The concave and convex parts which involve cutting grooves, and the making of protrusions call for precision, so I had an employee at the shop cut them for me. The Oldham's coupling that I saw in the museum was made from metal, but a wooden model was sufficient to reproduce the functionality.

Figure 4: Home-made Model of Oldham's Coupling

I later had another meeting with Oldham's coupling, when it was introduced at a gathering. Kunio Sugahara from Osaka Kyoiku University taught me that it was possible to make a model out of paper if straws with different diameters are used. Knowing that mathematical ideas are reflected in industrial machinery, I was somehow delighted to have majored in mathematics. If the motion shown in Figure 4 is difficult to follow, you can search for it on the internet and watch an MPEG format animation of Oldham's coupling. I found such animations on two or three sites. It seems that Oldham's coupling is being taught at colleges and in lectures on mechanical engineering.

3. Compressors in Air Conditioners

When I first presented this article in *Mathematics Seminar*, I received letters from many readers [2]. One of these letters revealed to me that "there is an interesting use of Oldham's coupling." This was referring to the use of Oldham's coupling in air conditioners. The principle behind the cooling performed by air conditioners is the compression and subsequent absorption of heat through the expansion of the gas. Air conditioners thus require compressors. These compressors were originally piston systems, but the technology advanced to rotary systems which utilize Reuleaux triangles, and apparently there have been further advances so that present models utilize so-called scroll systems. These scroll systems are not exactly Oldham's couplings, but they use the principle of the double-sliding crank. Scrolling systems produce little vibration or noise, and are also used in car air conditioners.

Why didn't we realize that mathematical ideas are used inside air conditioners? It's because the compressors are precision devices so they are set firmly by casting, and we cannot look inside them. Figure 5 shows that scroll systems are composed of a fixed scroll (in gray) and a mobile scroll (in black) which rotates around the fixed scroll in close contact with it. Gas drawn into the input port is compressed into the center and expelled through the discharge port within around 3 rotations of the mobile scroll.

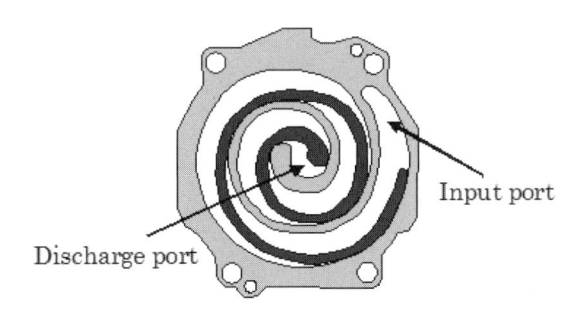

Figure 5: Compressor (Scrolling System)

Recently, so-called business museums have been growing popular, and there are many sites where businesses can introduce their products and permission is granted for general inspection at no charge. I visited the showroom of an air conditioner maker called *Daikin* in Osaka and was shown a model of the scroll system. The motion of this device is also interesting so I recorded an MPEG video using a digital camera. I was doubly surprised to see the link with Oldham's coupling that I had seen at Kyoto University Museum, and also to realize that this device was in use inside air conditioners.

4. The Involute Curves in Cogs

The curves used in the fixed and mobile scrolls are shown in Figure 6. They are known as 'involute' curves, and they are similar to well-known spiral curves although a little different. The construction of the scroll system shown in Figure 5 can be well understood as a pair of involute curves aligned with an offset of 180° and set in motion in the manner of Oldham's coupling.

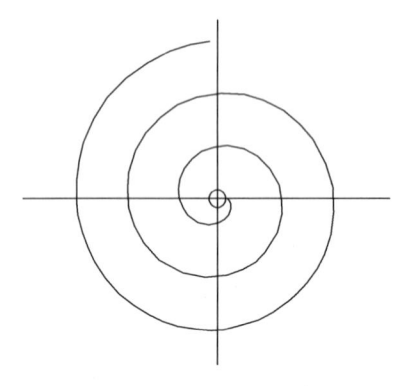

Figure 6: The Involute Curve

Involute curves were originally devised as a technical renovation of the shape of the teeth on cogs. They are the curve described by the tip of a thread while it is unravelled after having been wound around a circle (see Figure 7). When P is a point on the circumference of a circle O, the points $Q(x, y)$ at a distance $a\theta$ along the tangent from the point P are the coordinates of an involute curve. Expressing this as an equation yields

$$x = a\cos\theta + a\theta\sin\theta = a(\cos\theta + \theta\sin\theta)$$

$$y = a\sin\theta - a\theta\cos\theta = a(\sin\theta - \theta\cos\theta).$$

Cogs transmit motion between two rotating bodies, and the form of their teeth is particularly important. The shape has developed from cycloid or pin forms to the involute form, which is the most suitable form for cogs. The reason is illustrated in Figure 8. The teeth on cog O_1 are AA', and the teeth on cog O_2 are BB'. Suppose they are touching at point P. The contact point P moves from P_0 to P_3 along the mutual tangent line of the two cogs.

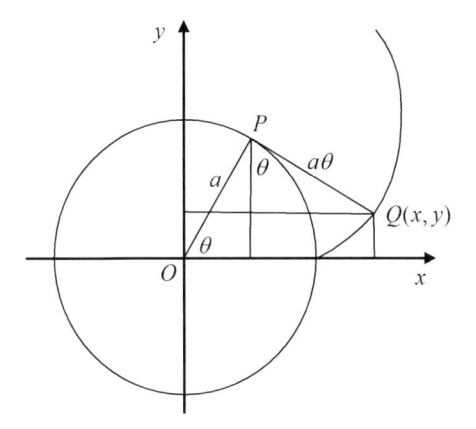

Figure 7: Cogs and Involute Curves

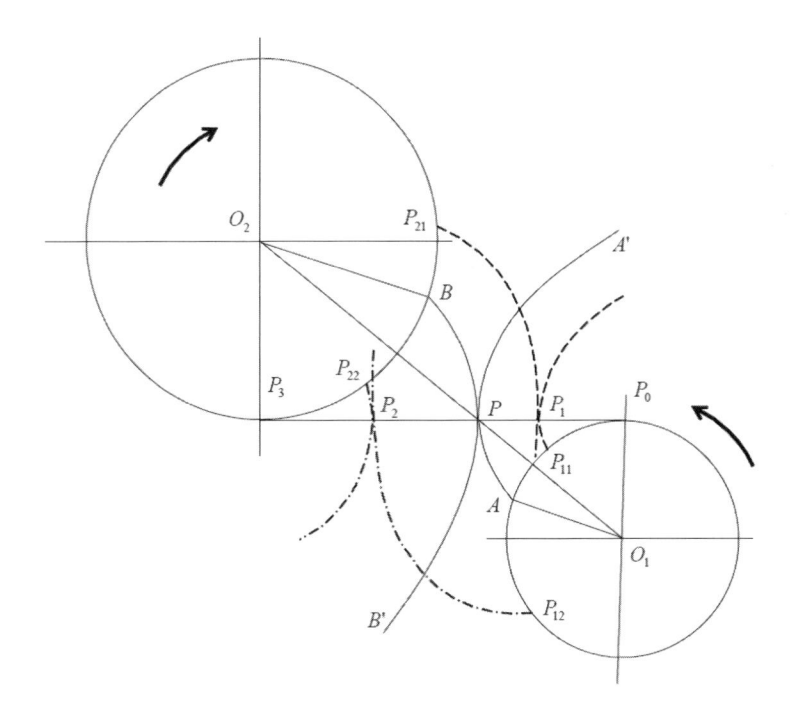

Figure 8: The Meshing of Involute-curve Cog Teeth

When O_1 rotates, the contact point advances to P_2 and the rotation is transmitted to O_2. At this point the form of the teeth is $P_{12}P_2$ and $P_{22}P_2$. Rotating O_1 backwards moves the contact point to P_1, and at that point the form of the teeth is $P_{11}P_1$ and $P_{21}P_1$. The rotational motion of both cogs is transmitted through the contact point. This contact point moves along the mutual tangent line. In order for the contact point to move along a straight line, the form of the teeth must be an involute curve. This is truly mathematical, isn't it? Indeed, mathematics doesn't just reside in textbooks, it is alive in our daily lives.

References

[1] H. Morita, *Kikogaku* [Mechanics], Tokyo: Jikkyo Press, 158-159, (1974).

[2] Y. Nishiyama, Hakubutsukan de mita Oldham Tsugite [A Discovery in a Museum: Oldham's Coupling], *Sugaku Semina* [*Mathematics Seminar*], 43(2), 7-9, (2004).

CHAPTER **18**
Measuring Areas: From Polygons to Land Maps

Abstract: This article explains how to measure area using both simple and complicated measurement methods. Elementary school children obtain area by counting triangles; junior-high school students use Cartesian coordinates; and high-school students study Heron's formula. In this article additional methods such as the trapezoid formula and Amsler's linear planimeter are presented.

AMS Subject Classification: 28A02, 00A09, 97A20
Key Words: Measurement of area, Polygon, Triangulation survey, Heron's formula, Trapezoid formula, Planimeter, Jacob Amsler, Polar planimeter, Linear planimeter

1. The Area of a Polygon

What kind of problem is this? Suppose we have an arbitrary polygon, where 'arbitrary' means that the number of vertices can be anything from 3 or more, and the shape does not have to be convex but can also include concave polygons. Let us assume that the number of vertices and the coordinates of each vertex are known.

$$P_1(x_1, y_1), P_2(x_2, y_2), \cdots, P_n(x_n, y_n)$$

Now let's try to obtain the area of such a polygon. Any method will do. But we'd like to have a simple algorithm that provides an accurate answer quickly, and ideally, is capable of handling an 'arbitrary' polygon. For example, how about the polygon shown in Figure 1, which has 8 vertices and is concave?

The most elementary method must surely be as follows. The 5th-year elementary school arithmetic syllabus includes the concept that "the areas of squares, pentagons, hexagons and so on can be obtained by dividing them up into a number of triangles." The octagon in Figure

1 can thus be divided up into 6 triangles (Figure 2). Then the area of each triangle can be obtained using the formula "base × height ÷ 2" and the total sum calculated. The line segment indicating the height of each triangle can be drawn nicely with a pair of set-squares, and measured with a ruler. The result will probably not be an exact number and will include some decimals, but it is certainly acceptable for elementary school pupils if they can follow this method, since they can add and multiply decimals.

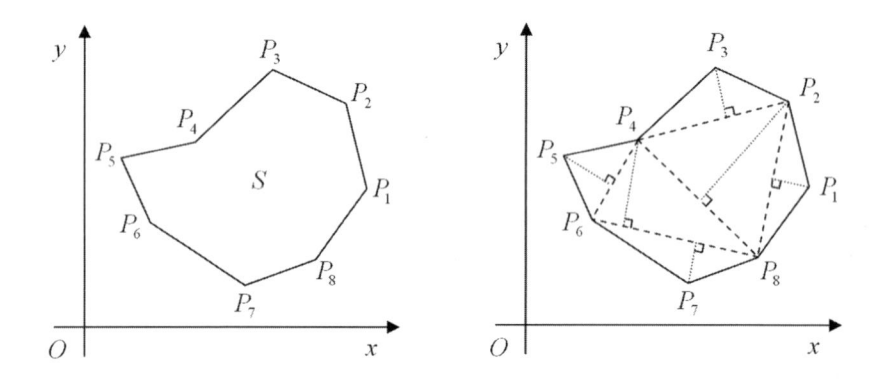

Figure 1: Find the Area!

Figure 2: Partitioning into Triangles

This most simple method is in fact used for measuring land even today, as the 'triangulation survey' method. However, this is a little too simple so we have another method we can apply. When students reach junior high-school they are expected to work on a larger number of problems on graph paper and they develop an awareness of Cartesian coordinates, so they can draw axes and auxiliary lines beside the axes, like those shown in Figure 3. As can be seen at a glance, the area of this polygon can be obtained by subtracting the areas of 8 triangles and 4 rectangles from the area of the outer rectangle. This method does not require measurements with a ruler, and can be obtained simply from the vertex coordinates.

2. Heron's Formula

The method described above does not incorporate a notion of generality regarding the 'arbitrary' aspect of an arbitrary polygon. For each different case it's necessary to try to draw auxiliary lines suited to the given diagram. Let's see if we can add generality to this calculation algorithm. Partition the polygon into 6 triangles by drawing auxiliary lines starting with vertex P_1, and passing through each vertex one-by-one from P_3 to P_7 (Figure 4).

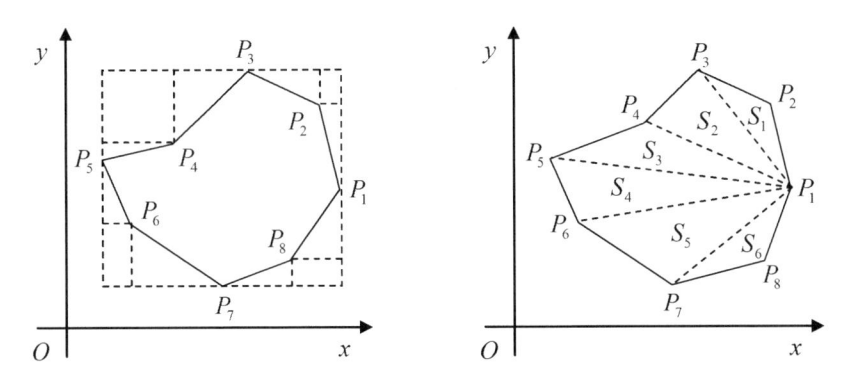

Figure 3: Parallel Auxiliary Lines   Figure 4: Generalized Partitioning

In general, an n sided convex polygon will be partitioned into $n-2$ triangles. When it comes to finding the area of the triangles, there is an equation known as Heron's formula. Although Heron's formula has currently been dropped from the high-school mathematics teaching guidelines, it is remarkably powerful. In short, taking the lengths of the three edges of a triangle as a, b, c, the area is

$$S = \sqrt{s(s-a)(s-b)(s-c)} \qquad (1)$$

(where $s = \dfrac{1}{2}(a+b+c)$).

The lengths of the edges can be obtained without measuring them using a ruler, by means of Pythagoras' theorem. Since the coordinates of each vertex are known, the distance $\overline{P_iP_j}$ between vertices $P_i(x_i, y_i)$ and $P_j(x_j, y_j)$ is

$$\overline{P_iP_j} = \sqrt{(x_i - x_j)^2 + (y_i - y_j)^2}. \qquad (2)$$

This is an effective method for adding the generality needed to handle n sided polygons. It's not necessary to construct each diagram individually. It is however a little bit tricky in cases when the diagram is too complicated to construct, and the presence of the square root makes paper-and-pencil calculation difficult.

3. Applying the Trapezoid Formula

So far, I have explained three methods, but isn't there a method for accurately obtaining the result, which is also simple and quick? The method I will now explain is not taught in high-school mathematics, though it is a rather elegant solution which uses the trapezoid formula. Regardless of whether or not the number of vertices is increased, or whether the polygon is concave, the result can be obtained very quickly using this method.

The principle is simple. A vertical line is dropped down to the x axis from each vertex. These 'legs' are denoted by $H_1, \cdots, H_8$ (Figure 5). The adjacent vertices and their vertical lines form trapezoids. For example, for the vertices P_1 and P_2, and their perpendicular lines H_1 and H_2, a trapezoid $P_1 H_1 P_2 H_2$ is formed with upper-base $P_1 H_1$, lower-base $P_2 H_2$ and height $H_1 H_2$. These trapezoids face sideways, and there are 8 of them in total, *i.e.*, there is one trapezoid for each vertex.

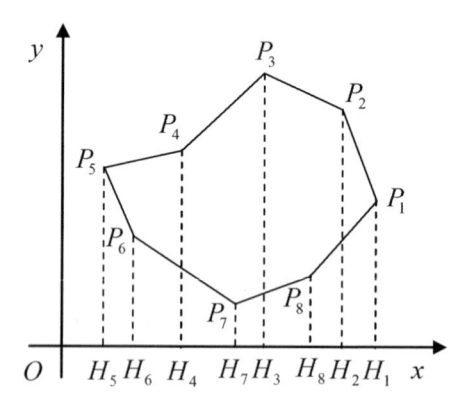

Figure 5: Dropping a Vertical Line Down from Each Vertex

The vertices are split into two groups, from P_1 to P_5, and from P_5 to P_1, and are shown in the diagrams in Figures 6 and 7. The areas of each of these trapezoids can be obtained with "(upper-base + lower-base) × height ÷2."

Let's try expressing the areas S_1 to S_8 using coordinate values. The areas S_1 to S_4 in Figure 6 are

$$S_1 = (y_1 + y_2) \times (x_1 - x_2) \div 2,$$
$$S_2 = (y_2 + y_3) \times (x_2 - x_3) \div 2,$$
$$S_3 = (y_3 + y_4) \times (x_3 - x_4) \div 2,$$

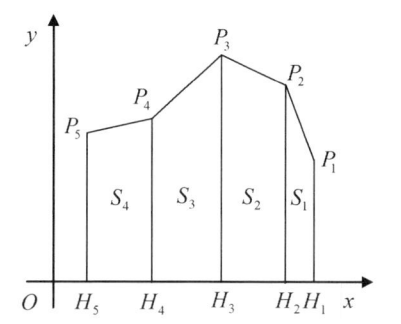

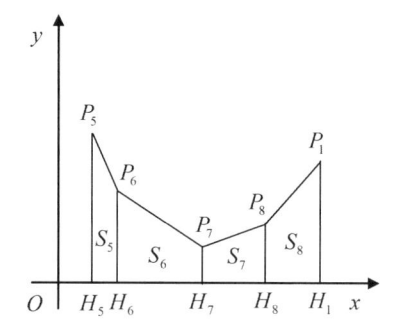

Figure 6: The Positive Area   Figure 7: The Negative Area

$$S_4 = (y_4 + y_5) \times (x_4 - x_5) \div 2. \quad (3)$$

The areas S_5 to S_8 in Figure 7 are

$$S_5 = (y_5 + y_6) \times (x_6 - x_5) \div 2,$$
$$S_6 = (y_6 + y_7) \times (x_7 - x_6) \div 2,$$
$$S_7 = (y_7 + y_8) \times (x_8 - x_7) \div 2,$$
$$S_8 = (y_8 + y_1) \times (x_1 - x_8) \div 2. \quad (4)$$

As can be understood from Figures 6 and 7, the area of the polygon is

$$S = (S_1 + S_2 + S_3 + S_4) - (S_5 + S_6 + S_7 + S_8). \quad (5)$$

Paying attention to the positively marked area in Figure 6, and the negatively marked area in Figure 7, the area of the polygon itself, S, can be composed as follows.

$$S = \frac{1}{2}\{(y_1 + y_2) \times (x_1 - x_2) + (y_2 + y_3) \times (x_2 - x_3)$$
$$+ (y_3 + y_4) \times (x_3 - x_4) + (y_4 + y_5) \times (x_4 - x_5)$$
$$+ (y_5 + y_6) \times (x_5 - x_6) + (y_6 + y_7) \times (x_6 - x_7)$$
$$+ (y_7 + y_8) \times (x_7 - x_8) + (y_8 + y_1) \times (x_8 - x_1)\} \quad (6)$$

In this equation, the reason why all the terms on the right-hand side are added is because the heights themselves are expressed with a sign, i.e., from S_1 to S_4, the heights are positive, but from S_5 to S_8 the heights are negative, and the corresponding areas themselves are also positive and negative.

In general, the area of an n sided polygon is

$$S = \frac{1}{2}\{(y_1+y_2)(x_1-x_2)+(y_2+y_3)(x_2-x_3)+\cdots+(y_n+y_1)(x_n-x_1)\}. \quad (7)$$

Since the subscripts on the parameters cycle, the program for finding the area is simple. Check for yourself how this works out in Visual Basic or C.

High-school mathematics textbooks contain the following explanation regarding integration. The area enclosed by two curves $y = f(x)$ and $y = g(x)$ is

$$S = \int_a^b \{f(x) - g(x)\}dx. \quad (8)$$

In the case of the area of a polygon, this is decomposed into

$$S = \int_a^b f(x)dx - \int_a^b g(x)dx, \quad (8)'$$

and each integral is handled using the trapezoid rule.

4. Areas on Maps

Areas on maps are not as easy to calculate as those of polygons. Terrain is expressed not with polygonal lines, but complicated curves. Polygonal lines with many vertices can be used to approximate curved lines but the result is not realistic. One frequently used method is to overlay square sections, and count the number of sections in order to find the area. It is also possible to cut out a paper shape of the terrain using a uniform material, measure the weight on a scale and thus obtain its area.

However, I'd like to introduce a more convenient method. It involves an instrument for measuring area known as a planimeter. The planimeter, which was devised in 1856 by the Swiss mathematician Jacob Amsler (1823-1912), is often used by architectural offices even now and it has a perfectly mathematical basis.

The basic construction of the planimeter is shown in Figure 8. The rods BA and BO are attached in such a way that they can freely rotate around a point. Given a closed curve denoted Γ which encloses the area we want to measure, an observation point O outside the curve is fixed.

Next, the survey point A follows the curve Γ in a clockwise circle. The number of rotations of the attached measurement wheel is read off, and the area can be then be obtained by multiplying this with the length of the pole. Although it may seem like some kind of magic that the area of the closed curve can be found with a single loop, it is in fact similar to the calculation of polygonal areas using trapezoids.

Figure 9 contains a diagram showing that the resulting area can be obtained when the point A follows a loop along Γ, and the area swept by the rod AB is deducted.

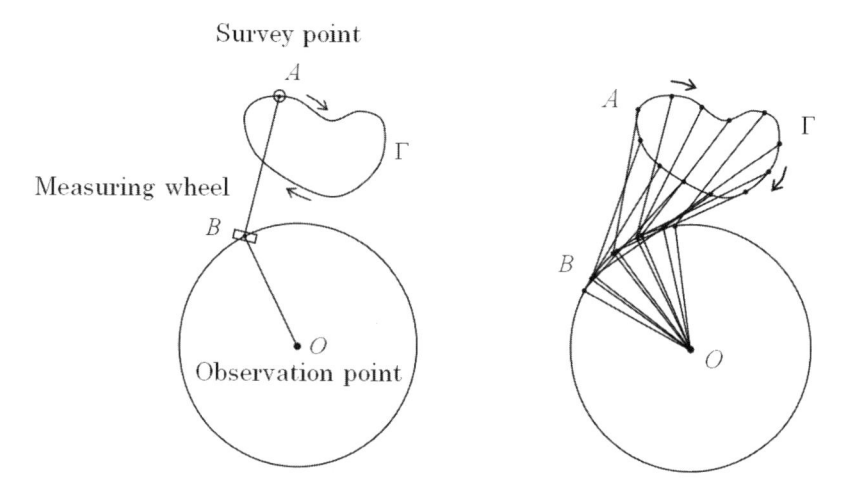

Figure 8: Reading Off the Number of Rotations of the Planimeter's Measuring Wheel

Figure 9: Pruning the Area through Which the Rod AB Extends

5. The Principle behind the Planimeter

The planimeter is a rather interesting instrument, so I'd like to briefly touch upon the principle behind it. Suppose that when the survey point A moves to A', the measuring wheel moves from B to B'. In general, a movement AA' can be separated into a translation $\overline{AA''}$ and a rotation $\overset{\frown}{A''A'}$. The area dA covered when the rod AB with length l translates by an amount ds and rotates by an amount $d\theta$ is

$$dA = lds + l^2 d\theta/2. \quad (9)$$

Supposing the measuring wheel tracking B rotates by dn, then

$$dn = ds. \qquad (10)$$

Substituting (10) into (9),

$$dA = l\,dn + l^2 d\theta/2. \qquad (9)'$$

At this point I'd like to add a little explanation regarding Equation (10). Figure 10 shows the relationship between $\angle ABO$ and the measuring wheel. Using α to denote the angle formed between the direction in which the measuring wheel rotates and the direction that the rod AB advances, the amount of rotation dn obeys the following relationship.

$$dn = \overparen{BB'} \cos\alpha \qquad (11)$$

Thus in the case of Figure 10(1), $\alpha = 0°$ and the measuring wheel has made a complete rotation. In case (3) $\alpha = 90°$, so the measuring wheel has not rotated at all. Case (2) is between cases (1) and (3), so the measuring wheel is sliding while it rotates, and has only rotated by $\cos\alpha$.

Now, the area is enclosed by one complete cycle of the closed curve so by integrating Equation (9)' we obtain the following formula,

$$S = \oint dA = \oint l\,dn + \oint \frac{l^2}{2} d\theta$$
$$= l \times n \qquad (12)$$

where n is the number of complete rotations. Also, the integral related to the rotational displacement $d\theta$ is zero for a full cycle, so

$$\oint d\theta = 0, \qquad (13)$$

and the 2nd term disappears. The number of rotations of the measuring wheel is thus proportional to the area.

The planimeter is an interesting instrument and I wanted to investigate it, but buying one can cost tens of thousands of yen. Luckily I managed to pick up an old planimeter much more cheaply through an internet auction. The device shown in Figure 11 is known as a 'polar planimeter' and it operates like that shown in Figure 8. The fixed observation point O is placed outside the map. Then the survey point A is looped around Japan's Lake Biwa in a clockwise fashion, the number of rotations made by the measuring wheel attached to the rod l is then read off, multiplied by the length of the rod l, and keeping the map's

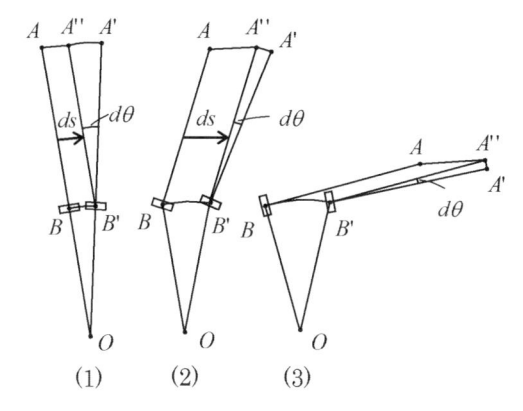

Figure 10: The Relationship between Angle ABO and the Measuring Wheel

scale of reduction in mind, the area can thus be obtained. The area of Lake Biwa was obtained with surprising accuracy.

It is also a surprise that this equation is linear. The device shown in Figure 12 does not have an observation point. The area is measured as the planimeter traverses the map surface. The survey point has become a lens looking down from above, and the measuring wheel is attached to the rod. There are wheels attached to both the left and right ends of the rod, which can only move backwards and forwards in a region with a fixed width. This formula is often used today. There are other different formulae for planimeters, but all of them embody mathematical principles.

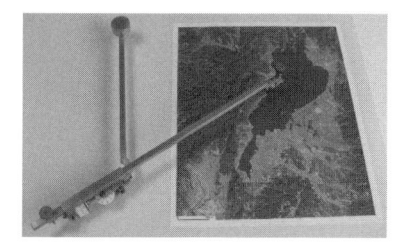

Figure 11: Polar Planimeter   Figure 12: Linear Planimeter

CHAPTER **19**

Sicherman Dice: Equivalent Sums with a Pair of Dice

Abstract: George Sicherman discovered an interesting pair of dice whose sums have the same probability distribution as a pair of standard dice, and this was reported by Martin Gardner in 1978. This pair of dice is numbered 1, 3, 4, 5, 6, 8 and 1, 2, 2, 3, 3, 4, and is unique. In order to prove the uniqueness of his combination, three methods are shown: trial-and-error with pencil and paper, a Visual Basic program and factorization of polynomials. The third is the most elegant solution and was presented by Gallian and Rusin, as well as Broline in 1979.

AMS Subject Classification: 11A02, 00A09, 97A20
Key Words: Sicherman dice, Probability distribution, Generating function, Factorization of polynomials

1. One Particular Problem in Probability

I received a rather interesting mail from Dr. Steve Humble, an English acquaintance from the International Congress on Mathematical Education (ICME10) held in Denmark in the summer of 2004. It was regarding a problem in probability which involves finding the sum of the numbers given by two dice. The sums of the numbers on the dice are distributed from 2 to 12, and if we had a different pair of dice with the same probability distribution, they would be marked with the numbers
$$1, 3, 4, 5, 6, 8 \text{ and } 1, 2, 2, 3, 3, 4.$$
The problem, in particular, was to prove that this is the unique solution (besides the pair of standard dice both marked 1, 2, 3, 4, 5, 6).

Problems involving the probability distributions of dice often come up in exams. For example, 'what is the probability that throwing two dice yields a sum which is even,' '...that the sum is a multiple of 3,' '...that the sum is greater than 5?' *etc.* These questions assume standard dice with the numbers from 1 to 6 on their faces. I read the mail from my acquaintance, thinking to myself that this was another such problem, and tried to confirm the calculation regarding the sums of the numbers on the faces.

Dice are cubes, so they have 6 faces. Taking two standard dice and writing out the frequency distribution of their sum yields the results shown in Figure 1. There are 6 different outcomes from the first die, and another 6 from the second die, giving a total set of 36 outcomes. The sum of 2 occurs 1 time, the sum of 3 occurs 2 times, ..., the sum of 7 occurs 6 times, ..., and the sum of 12 occurs 1 time. The sum of 7 has the most occurrences and the frequency distribution graph has a triangular shape which increases and decreases linearly.

Exchanging the numbers on the first and second dice with the numbers that I received by mail, 1, 3, 4, 5, 6, 8 and 1, 2, 2, 3, 3, 4, and computing the sum of the two dice using some spreadsheet software, I could determine the frequency distribution. What was the result? The sum of the numbers is between 2 and 12, and somehow the frequency distribution is the same as that of normal dice as was shown in Figure 1 (see Figure 2).

		\multicolumn{6}{c}{Die 2}
		1	2	3	4	5	6
Die 1	1	2	3	4	5	6	7
	2	3	4	5	6	7	8
	3	4	5	6	7	8	9
	4	5	6	7	8	9	10
	5	6	7	8	9	10	11
	6	7	8	9	10	11	12

		\multicolumn{6}{c}{Die 2}
		1	2	2	3	3	4
Die 1	1	2	3	3	4	4	5
	3	4	5	5	6	6	7
	4	5	6	6	7	7	8
	5	6	7	7	8	8	9
	6	7	8	8	9	9	10
	8	9	10	10	11	11	12

Figure 1: Normal Dice

Figure 2: Dice with the Same Probability Distribution as Normal Dice

I was immediately impressed by this, but then wondered whether it is just the result of chance, and tried to think whether there were some other sets of dice with the same probability distribution. Since the sum of 2 occurs once with normal dice, it is immediately clear that the sum of each die's smallest value must be 2. Writing an x for the numbers which are not yet determined yields the following.

$1, x, x, x, x, x$ and $1, x, x, x, x, x$

In addition, the largest sum of 12 also occurs only once, so adding up the last numbers must result in 12. This results in the following cases.

$1, x, x, x, x, 10$ and $1, x, x, x, x, 2$

$1, x, x, x, x, 9$ and $1, x, x, x, x, 3$

$1, x, x, x, x, 8$ and $1, x, x, x, x, 4$

$1, x, x, x, x, 7$ and $1, x, x, x, x, 5$

It is now sufficient to patiently and mechanically investigate the remaining numbers marked x. Of the 4 x marks in the middle, the smallest value on the left cannot include 1, so it must be at least 2. Also, the large number on the right must not include the largest number so it must be less than this value. It is not therefore necessary to try and find a set of numbers at random, but rather, by thinking about the conditions and only selecting numbers that satisfy the conditions, the problem ceases to be a headache and the solution can be obtained with just pencil and paper.

So with just a little investigation it became clear that the first 2 of the 4 cases shown above were not possible. The numbers in the 4th case are good in one sense, but they are not perfect. For example, in the case of 1, 2, 4, 4, 6, 7 and 1, 2, 3, 3, 4, 5, the frequencies of the sums 5, 6, 8, 9 are out by 1. This means that any solution must exist in case 3, which is $1, x, x, x, x, 8$ and $1, x, x, x, x, 4$. This result was obtained using only pencil and paper.

Taking a look at the sets of numbers in the correct solution 1, 3, 4, 5, 6, 8 and 1, 2, 2, 3, 3, 4, their averages are 4.5 and 2.5, respectively. The sum of these averages is 7. The distribution of these numbers is also symmetric from left to right in the shape of a mountain. Standard dice have an average value of 3.5 and the sum of the averages is 7. Aren't the facts that sums of the average values are equal, and that the numbers are distributed in a well-balanced way, essential conditions for the probability distributions to be the same?

While we might not be able to describe this is an elegant solution, we were able to find the dice with the same frequency distributions by ourselves. However, this was a trial-and-error method using paper and pencil, so it's possible that there could have been an oversight. Also, it's hard to say that the method of proof that I thought of is mathematical.

Was the existence of one more set of dice with the same probability distribution chance or necessity? The first round of this investigation ends with various questions left unanswered.

2. For Octahedral Dice, There are 3 Solutions

I didn't think I would want to write a program when I was abroad, so I didn't take my computer software with me. However, to check this problem I did, in fact, need software. This is because there is a high probability of making a mistake with a calculation by hand. I obtained Visual Basic cheaply through an internet auction, installed it, made a simple program and performed the check. If the program is assembled thoughtlessly it takes a lot of computation time and no result is generated. For example, for a single die, choosing the 6 numbers requires 6 loops, and since there are two dice this must be performed twice, so there a total of 12 loops. It is then necessary to investigate the 6 possible occurrences of the numbers between 1 and 6. It is sufficient to investigate every single one of these values, but it's not a very good method.

Obtaining the paper and pencil solution before writing the program turned out to be a preliminary investigation for writing an efficient program. There are 6 numbers to decide for the dice, but the first number of 1 and the last number 'max' are fixed so it is sufficient to investigate only those 4 intermediate numbers. Also, the numbers are in ascending order from left to right, so it's not necessary to investigate all possible numbers. If attention is paid to these points the program can obtain a solution efficiently. Executing a program constructed in this way reveals the result that the only set of dice with the same probability distribution for the sum are those with the numbers 1, 3, 4, 5, 6, 8 and 1, 2, 2, 3, 3, 4. There were no mistakes in calculation due to arithmetic by hand, or oversights.

Satisfied by the result of the program, my interest developed in a different direction. Was it merely a coincidence that 1, 3, 4, 5, 6, 8 and 1, 2, 2, 3, 3, 4 is the only solution? Dice are regular 6-faced bodies. The other regular polyhedra besides this are the tetrahedron with 4 faces, the octahedron with 8 faces, the dodecahedron with 12 faces and the icosahedron with 20 faces. Wouldn't it be possible to set up similar probability problems by considering these other regular polyhedra as dice? With this in mind, I thought about the octahedron which has 2 extra faces. In the case of 'standard' octahedra, the numbers on the dice are

$$1, 2, 3, 4, 5, 6, 7, 8 \text{ and } 1, 2, 3, 4, 5, 6, 7, 8,$$

so the sum is distributed between 2 and 16, and the sum with the highest frequency is 9, which occurs 8 times. Is there another set of numbers for the octahedron with the same probability distribution, or perhaps not?

Maybe not only one solution can be found, but many? I tried to make a prediction. For cubes there is only one case which is a solution. Since the number of sets of resulting numbers is proportional to the product of the number of faces on the dice, with regular octahedra there are more than with the cube, and I wondered if more complicated cases might be found. That is to say, the range to search is not $6 \times 6 = 36$ cases but rather $8 \times 8 = 64$. Doesn't this increase the chances of finding a solution?

I completed a suitable program with just a few modifications to the program used to check the cubes. It was easy to make a simple change increasing the number of faces to investigate from 6 to 8.

Running it revealed that my prediction was correct. The following 3 solutions were found for the case of regular octahedra,

$1, 3, 3, 5, 5, 7, 7, 9$ and $1, 2, 2, 3, 5, 6, 6, 7$,
$1, 2, 5, 5, 6, 6, 9, 10$ and $1, 2, 3, 3, 4, 4, 5, 6$,
$1, 3, 5, 5, 7, 7, 9, 11$ and $1, 2, 2, 3, 3, 4, 4, 5$.

I suggest you check for yourself that these sets of numbers have the same probability distributions as normal octagonal dice. There was only 1 solution for the case of cubes, but in the case of octahedra there are 3. It may be thought that the likelihood of finding solutions is proportional to the product of the number of faces after all.

Among the regular polyhedra, there are also dodecahedrons (12 faces) and icosahedra (20 faces). I therefore tried to investigate dodecahedra. However, in order to investigate this using my program, the number of faces must be increased from 8 to 12, and this resulted in a massive increase in computing time. I realized that it was impossible to find a solution using this method and gave up. I found out later, by means of another method I discovered, that there are 7 solutions for the case of the 12-faced dodecahedron. For the 20-faced icosahedron it is easy to predict that this will be increased even further. Compiling the results above, for the 6-faced cube there is 1 solution, for the 8-faced octahedron there are 3 solutions, and for the 12-faced dodecahedron there are 7 solutions. Plotting this data with the number of faces along the horizontal axis and the number of solutions on the vertical axis reveals that the shape is not linear but rather a 2nd or 3rd order curve.

Giving up on the regular polyhedron with the most faces, I instead investigated the 4-faced tetrahedron. It was unexpected, but I also found 1 solution for tetrahedra. It is the set

$1, 3, 3, 5$ and $1, 2, 2, 3$.

So, I thought to myself, a solution with the same probability distribution exists for tetrahedra. I felt impressed and also satisfied, having achieved some positive results by checking with my Visual Basic program. This check using the program makes no slip-ups as might occur when calculating by hand, and is a convenient method of investigation. However, perhaps because I had a sense of having relied on something other than myself to solve the problem, I didn't feel like I had achieved the solution and I was left feeling somewhat dissatisfied.

3. Proof using Polynomials

I organized the results of my program, including the fact that with octagonal dice there are three sets of numbers with the same probability distribution, and mailed them to Dr Steve Humble. Apparently he didn't already know these results. In reply to my mail he explained the following to me. Regarding a proof, this probability problem appears in an old work by Martin Gardner, which involves something known as a generating function. It has the following form.

$$P(x) = \frac{1}{6}(x + x^2 + x^3 + x^4 + x^5 + x^6) \quad (1)$$

It is possible to make a proof using this formula.

Martin Gardner was a mathematics essayist who was active during the 1970s. He had many readers and was very influential. Some more senior readers may already know the conclusion of this problem. I am poor at probability so I have consciously avoided probability problems, but for some reason I found it interesting to take this one up. The proof is stated below, but it was established in the 1970s so there has been 30 years of progress up to the present day. The mode of expression has changed a little, but it is roughly the same as the original method, which is based on the use of a polynomial. Allow me to explain an outline of the proof.

The following point describes how to read Equation (1), since it is not the type of problem that involves substituting a value for x and obtaining the value of the equation. The proof uses the exponents and coefficients of the polynomial, so I'd like you to get used to paying attention to the exponents and coefficients. There are a total of 6 terms, and each has the form ax^k which is read as meaning that there are a cases when the sum of the face values is k. The exponent is the sum of the face values, and the coefficient corresponds to the frequency. Let's take a look at a concrete example. For a single die there is 1 case when the face value is 1, 1 case when the face value is 2, ..., and 1 case when the face value is 6. This yields $1x^1, 1x^2, \cdots, 1x^6$, i.e., $x, x^2, \cdots, x^6$. Each of the cases occurs with the same probability, and since the overall sum of all the probabilities must be 1, it is divided by 6.

The case of two dice corresponds to squaring the left- and right-hand sides of Equation (1). Then, when the resulting polynomial is expanded, the exponents and coefficients express the sums of the face values and their frequencies, respectively. The following mathematical expression affirms the above. In order to improve the readability, both sides of Equation (1) are multiplied by 6, thus removing the denominators.

$$6P(x) = x + x^2 + x^3 + x^4 + x^5 + x^6 \quad (1)'$$

$$\{6P(x)\}^2 = x^2+2x^3+3x^4+4x^5+5x^6+6x^7+5x^8+4x^9+3x^{10}+2x^{11}+x^{12} \quad (2)$$

Paying attention to Equation (2), it can be seen that the exponent and coefficient of every term in the expanded polynomial neatly expresses the relationship between the sums of the face values and their frequencies, and the probability distribution. For example, $5x^6$ is read as meaning that there are 5 cases when the sum of the face values is 6, and $4x^9$ means there are 4 cases when the sum is 9. What is needed to solve this problem is to identify what forms the right-hand side of Equation (2) can take if it can be factorized. If the factorization takes the form of the polynomial in Equation (1), then doesn't this correspond to a normal die? This is not true. If it could be factorized as the product of different polynomials, then this constitutes a solution. But can this be done?

Let's return to Equation (1)' and try transforming it as follows. Each term includes the factor x, so first factorize by x and gather the remaining terms, $(1 + x + x^2 + x^3 + x^4 + x^5)$. As is well known, multiplying $\sum_{i=0}^{k-1} x^i$ by $(x-1)$ yields $(x^k - 1)$, so let's apply this fact. Multiplying both the numerator and the denominator by $(x-1)$ does not change the value of the formula. In general, the form $(x^k - 1)$ is easy to factorize, and as a consequence many factors can be found, e.g., $(x^6 - 1)$ yields $(x-1)(x^2+x+1)(x+1)(x^2-x+1)$.

Putting the above facts together and writing out the result yields the following.

$$6P(x) = x(x^5 + x^4 + x^3 + x^2 + x + 1)$$
$$= x(x^5 + x^4 + x^3 + x^2 + x + 1)(x-1)/(x-1)$$
$$= x(x^6 - 1)/(x-1)$$
$$= x(x^3 - 1)(x^3 + 1)/(x-1)$$
$$= x(x-1)(x^2+x+1)(x+1)(x^2-x+1)/(x-1)$$

$$= x(x^2 + x + 1)(x + 1)(x^2 - x + 1) \tag{3}$$

As shown in Equation (3), it was possible to factorize $P(x)$ into a product of four terms. This is a form that one wouldn't imagine from looking at Equation (1). Simply exponentiating the left- and right-hand sides of Equation (3) results in the following form.

$$\{6P(x)\}^2 = x^2(x^2 + x + 1)^2(x + 1)^2(x^2 - x + 1)^2 \tag{4}$$

The point is to solve the problem by recomposing Equation (4) into two parts, but it cannot just be partitioned at random. After exponentiating, the total number of terms is 8. There are conditions on the partitioning, so let's investigate them. Returning to Equation (3) and investigating each term reveals firstly that each partition must contain x because this corresponds to the faces of the dice marked 1. Regarding the 3 remaining terms, examining the value in the case when $x = 1$ reveals the following.

$(x^2 + x + 1) = 3$
$(x + 1) = 2$
$(x^2 - x + 1) = 1$

The left-hand side of $P(x)$ is multiplied by 6, so unless the right-hand side also contains the factor 6 then the equation is not balanced. Because of this, the 3 from the term $(x^2 + x + 1)$ and the 2 from the term $(x + 1)$ must each be included once. With $3 \times 2 = 6$, the equation is balanced with respect to the number 6 and passes this requirement. On the other hand, the value of $(x^2 - x + 1)$ is 1, so it has no effect, regardless of whether it is included or not. As for the way the $(x^2 - x + 1)^2$ terms are partitioned, if one is included in each die, this must be the same as the case of normal dice. If they are included on only one side, it results in dice that differ, which is the solution we were trying to obtain.

The result of this investigation is shown in Equation (5). The left-hand side means 2 normal $P(x)$ dice, and the right-hand side means a set of differing $Q(x)$ and $R(x)$ dice. Expanding the products of each polynomial, they are the same as the right-hand side shown in Equation (2), *i.e.*, the probability distributions of the sums of the face values are the same.

$$\{6P(x)\}^2 = \{6Q(x)\}\{6R(x)\}$$
$$= \{x(x^2 + x + 1)(x + 1)(x^2 - x + 1)^2\}\{x(x^2 + x + 1)(x + 1)\}$$
$$= (x^8 + x^6 + x^5 + x^4 + x^3 + x)(x^4 + 2x^3 + 2x^2 + x) \tag{5}$$

The polynomials of the solution are:

$$Q(x) = \frac{1}{6}(x^8 + x^6 + x^5 + x^4 + x^3 + x) \tag{6}$$

$$R(x) = \frac{1}{6}(x^4 + 2x^3 + 2x^2 + x) \tag{7}$$

Since the exponents and coefficients of each of the terms in the polynomials express the sums of the face values and their frequencies, die Q has the numbers 8, 6, 5, 4, 3 and 1 on its faces, and die R has 4, 3, 3, 2, 2 and 1 on its faces.

These numbers are the same as those of the unique solution for cubes presented at the beginning. This method of proving the result by utilizing polynomial exponents and coefficients is nothing short of brilliant. The explanation presented dealt with the proof for cubes, but this method can be applied to the proofs for tetrahedra, octahedra, dodecahedra and icosahedra. Because of the limits on computational speed when finding solutions using the Visual Basic program, it was not possible to obtain solutions above dodecahedra, but it is possible using the polynomial method. The discovery of the 7 solutions in the case of the dodecahedron was also due to the polynomial method.

4. Sicherman Dice

Having achieved an understanding of the proof based on polynomials the matter is settled. Who thought of this interesting problem, and who realized there was such an elegant solution method? My interest shifted to searching for the roots of this problem and solution method.

As a result of various research, I was able to track a trail leading back to the 1970s. The problem's first appearance seems to have been in the February 1978 edition of *Scientific American*, which carried an article by Gardner on page 19 [3]. Reading this article reveals that the person who initially discovered these peculiar dice was George Sicherman. It is not certain whether or not he knew a proof. He probably just presented the fact that these interesting dice exist. Many letters concerning proofs arrived for Gardner from readers who read this journal, and Gardner later wrote that the elegant solution method using a polynomial representation was owing to J.A. Gallian and D.M. Broline. Two relevant papers are given in the reference section [1],[2].

These days the dice are known as 'crazy dice' or take the name of their discoverer, *i.e.*, 'Sicherman dice.' Some companies (not Sicherman) sell these dice as merchandise, although it seems that they are not used in actual casinos. This means that while the issue of equivalent sums with a pair of dice is a fascinating topic for mathematicians, it is a different matter in reality!

References

[1] D.M. Broline, Renumbering of the faces of dice, *Mathematics Magazine*, 52(1979), 312-315, (1979).

[2] J.A. Gallian, D.J. Rusin, Cyclotomic polynomials and nonstandard dice, *Discrete Mathematics*, 27(1979), 245-259, (1979).

[3] M. Gardner, Mathematical Games, *Scientific American*, 238(1978), 19-32, (1978).

CHAPTER **20**

Unexpected Probabilities

Abstract: This article presents two unexpected probabilities. If there are at least 23 people in a class, then the probability that there is a shared birthday exceeds one half. This is explained by the idea of complementary events. What are the chances of there being a student who ends up in the same seat when everyone in the class is reseated? This probability is around $\frac{2}{3}$ irrespective of the number of students, and was solved by Montmort (1708).

AMS Subject Classification: 60A02, 00A09, 97A20
Key Words: Likelihood of shared birthdays, Complementary events, Probability of ending up in the same seat, Montmort's theory, Venn diagram

1. The Likelihood of Shared Birthdays

There are all kinds of problems in probability, and the study of probability gets more and more interesting when there is a big gulf between expectation and reality. I refer to examples of this gulf between expectation and reality as 'unexpected probabilities,' and in this chapter I'd like to introduce two such probability problems.

What kind of value do you expect for the probability that two people in a class share the same birthday? There are about 40 people in one of my classes. I ask students to estimate the probability that 2 people have the same birthday, assuming there are 365 days in one year. Most students estimate on the side of 'no one.' This is because they reason that while there are 40 students in the class, there are 365 days in the year, so the chance of a shared birthday is $\frac{40}{365}$. They also suppose that the first time there would be students with the same birthday is with a class of 366 students.

One's understanding of probabilities cannot be deepened without trying out some practical experiments. I take up this example every year in my lectures on information mathematics intended for students in the humanities. I have students say their birthdays, and after hearing from all the students once, I ask students whose birthdays were duplicated to raise their hands. Enough students raise their hands at this point that it's safe to say there's always some. Owing to the fact that they estimated the probability as being around $\frac{40}{365}$, students find this considerable discrepancy mysterious. It is at this point that I provide a mathematical explanation of probability.

This is a problem which is solved using the concept of complementary events. The number of people in the class is taken to be n, and the number of days in one year is taken to be 365. First we obtain the probability that everyone in the class has a different birthday. There are 365 days from which the first person's birthday may be chosen, so the probability associated with the first person is $\frac{365}{365}$. There is then one less day in the year from which the second person's birthday may be chosen, leaving 364 days. The probability is $\frac{364}{365}$. For the nth person, the probability associated with their birthday is $\frac{365 - (n-1)}{365}$, so for a total of n people, the probability that they all have different birthdays is

$$P_1 = \frac{365}{365} \times \frac{365 - 1}{365} \times \cdots \times \frac{365 - (n-1)}{365}.$$

The probability that there is at least one shared birthday can be obtained as the probability of a complementary event as described above,

$$P = 1 - P_1 = 1 - \frac{365}{365} \times \frac{365 - 1}{365} \times \cdots \times \frac{365 - (n-1)}{365}.$$

Furthermore, when $n = 23$, we have $P = 0.507$. This means that if there are at least 23 people in the class then the probability that there is a shared birthday exceeds one half. Considering this value of 23, it is less than one tenth of the 365 days.

The concept of complementary events is extremely useful. The concepts in the phrases 'denying that all the birthdays are different' and 'at least one birthday is repeated' are expressed using the notions of 'not always' and 'at least,' respectively. Problems involving complementary events often come up, so it is surely important to gain familiarity with them. This kind of surprising probability problem is not widely known, and the reason it does not often appear in exams likely stems from the fact that it is difficult to calculate P_1 and P on paper. Despite being an interesting example, forgetting about it just because it is not suited to examinations is not ideal.

Personal computers are now prevalent, and by using spreadsheet software the calculation is rendered straightforward. Calculating the probability of there being a shared birthday, P, while varying the number of people in the class, n, and graphing the results must yield a deeper understanding of probability. When n is small ($n < 10$), P has a small value ($P < 0.1$), but beyond that it rapidly increases so that for $n = 23$ the probability is $P = 0.507$, and for $n > 40$, $P > 0.9$. This birthday-related example can usually be demonstrated practically as a class experiment, so it may be introduced to a class with confidence.

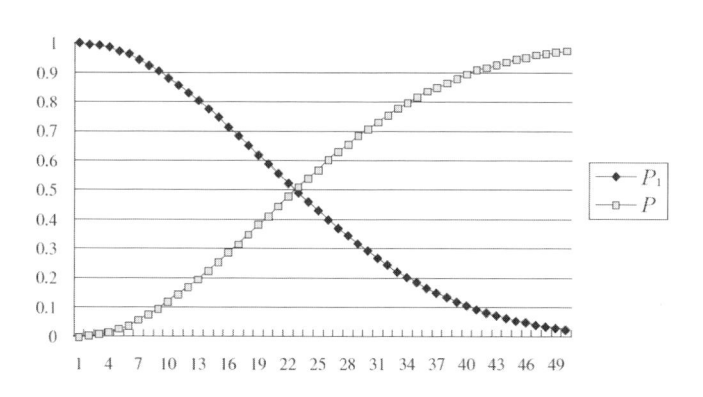

Figure 1: The Probability That There Is At Least One Shared Birthday in the Class

2. What's the Probability That At Least 2 Pairs Share a Birthday?

I lecture about the birthday example every year. In fact, this year when I performed the experiment with a class of 60 people, the result revealed 3 pairs with the same birthday. If there is at least 1 pair it is sufficient for the lesson, but there were 3 pairs who shared the same birthday. This gave rise to a new further enquiry as to whether the fact that among 60 people there were 3 pairs with the same birthday was an appropriate number. When I performed the experiment with a class 2 years ago, there were 40 students in the class, and there was 1 pair with the same birthday. But this year, for 60 students there were 3 pairs. Moreover, one of the 3 pairs was a pair of fellow students who sat next to each other. Is this an appropriate number after all?

For simulations it is convenient to use random numbers. A confirmation is possible using Visual BASIC, for which the built-in random-number generating function RND can be used with a program of about 50 lines. The chance of selecting a pair of people is given by the following formula.

$$_nC_2 = \frac{n!}{(n-2)!2!}$$

For each case, the remaining $(n-2)$ people's birthdays must fall on different days, which has the following probability.

$$P_2 = {_nC_2} \times \frac{365}{365} \times \frac{365-1}{365} \times \cdots \times \frac{365-(n-2)}{365}$$

The probability that 2 or more pairs of people share a multiple birthday, or more than 2 people share the same birthday is

$$1 - P_1 - P_2.$$

Calculating the equation above for $n = 23$ and $n = 60$ using spreadsheet software reveals that the values take the following form.

n	P_1	$1-P_1$	P_2	$1-P_1-P_2$
23	0.493	0.507	0.363	0.144
60	0.006	0.994	0.034	0.960

Table 1. Probabilities of Multiple Birthdays ($n = 23, 60$)

When calculating the probability P_2 using spreadsheet software, calculating the numerator and denominator separately as follows causes an overflow during computation.

$numerator = 365 \times {_nC_2} \times (365-1) \times \cdots \times (365-(n-2))$
$denominator = 365 \times 365 \times 365 \times \cdots \times 365 = 365^{n-1}$

The technique for preventing an overflow during the computation of P_2 is to compute each term completely, in sequence. Proceeding in this way reveals that for 60 people, 3 or 4 pairs with the same birthday is not at all unreasonable.

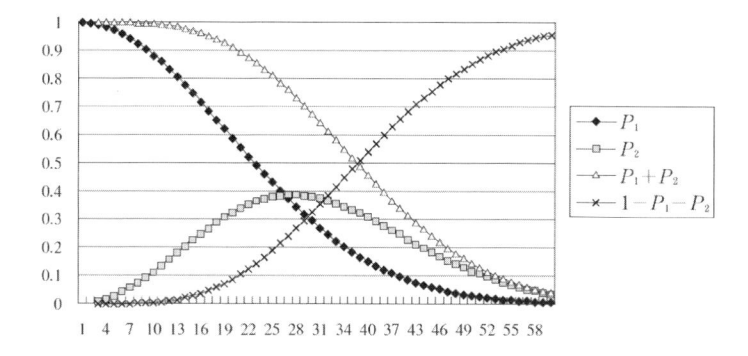

Figure 2: The Probabirity That At Least 2 Pairs Have the Same Birthday

3. The Probability of Ending Up in the Same Seat

I'd like to introduce another surprising probability. This is a probability associated with changing seats. What are the chances of there being a student who ends up in the same seat when everyone in the class is reseated? Isn't there someone who has had the unfortunate experience of not getting a different seat when everyone is reseated? Even if it wasn't yourself you must have known a friend who didn't get to change. Let's look at this and explain why it is by no means a great misfortune, but rather an event that occurs with a high probability.

The reseating problem is the same as the 'Secret Santa' Christmas present problem. Should we take the pessimistic perspective that it is a punishment for bad behavior when someone draws the same present that they brought themselves? This is not so, the probability is surprisingly high.

Before writing up a mathematical expression, let's attempt a counting method. Each of the students are denoted by a number 1, 2,··· and compared before and after the reseating. Cases where the seat is the same before and after reseating are marked with a circle. When there are 2 students there are 2 seats and the number of ways of arranging them is $2! = 2$. As shown in the first line of Table 2, students 1 and 2 both remain unchanged in 1 case. Since there is 1 case in which a student ends up in the same seat, the probability of this happening is 0.5.

Before reseating	1	2
After reseating	1*	2*
	2	1

Table 2. The Case of $n = 2$

In the same way, the case of 3 students is shown in Table 3. There are $3! = 6$ ways of arranging the students, and the cases when a student ends up in the same seat are indicated by asterisks. For the case in line 1, students 1, 2 and 3 all end up in the same seat. For the case shown in line 2, only student 1 ends up in the same seat. In line 3 only student 3, and in line 6, only student 2 ends up in the same seat. There are 4 cases in which there is a student whose seat does not change, so the probability of this happening is as follows.

$$P = \frac{4}{3!} = \frac{4}{6} = 0.666\cdots$$

Before reseating	1	2	3
After reseating	1*	2*	3*
	1*	3	2
	2	1	3*
	2	3	1
	3	1	2
	3	2*	1

Table 3. The Case of $n = 3$

[Question 1] Find the probability that at least one student ends up in the same place for $n = 4$, using a counting method.

There are $4! = 24$ ways of arranging the students after reseating, and there are 15 cases in which there is an unchanged seat. The probability of this happening is therefore

$$P = \frac{15}{4!} = \frac{15}{24} = \frac{5}{8} = 0.625.$$

This counting method is valid, but it's also important to be able to capture the problem mathematically. It is known that for n students, the probability that at least 1 person does not change their seat is

$$P = 1 - \frac{1}{2!} + \frac{1}{3!} - \frac{1}{4!} + \cdots \pm \frac{1}{n!}.$$

Let's think about the derivation of this formula.

When 1 student doesn't change his or her seat, the remaining $(n-1)$ students may be arranged in $(n-1)!$ ways. This works out equivalently whichever of the n students ends up in the same seat, and there are

$$n \times (n-1)! = n!$$

ways in total. In this case some of the remaining students might end up in their original seats, and this situation is discussed below.

In the case that student 1 and student 2 do not change their seats, the remaining $(n-2)$ students may be arranged in $(n-2)!$ ways. There are ${}_nC_2$ ways of choosing the 2 students, so there are

$$_nC_2 \times (n-2)! = \frac{n!}{(n-2)!2!} \times (n-2)! = \frac{n!}{2!}$$

ways in total.

In the same way, the case when 3 students such as student 1, student 2 and student 3 do not change their seats occurs in

$$_nC_3 \times (n-3)! = \frac{n!}{3!}$$

cases in total.

The relationship above is a little complicated, so for $n = 3$, let's use a Venn diagram to explain the situation (Figure 3). The event that student 1 does not change is denoted A_1, the event that student 2 does not change is denoted A_2, and for student 3, A_3. The event that at least 1 student does not change seat, A, is thus described as follows.

$$A = A_1 \cup A_2 \cup A_3 = (A_1 + A_2 + A_3) - (A_1 A_2 + A_2 A_3 + A_3 A_1) + A_1 A_2 A_3$$

For the event A_1, there are 4 cases, *i.e.*, when only student 1 doesn't change seat, when neither student 1 nor student 2 changes seat, when neither student 1 nor student 3 changes seat, and when neither student 1 nor student 2 nor student 3 changes seat. The equation above reflects the elimination of these overlapping cases. $(A_1 + A_2 + A_3)$ is a simple sum of all the 3 events. The parts that are over-included by duplication are removed by the term $(A_1 A_2 + A_2 A_3 + A_3 A_1)$. However, this again removes too much, so the term $A_1 A_2 A_3$ is added back in.

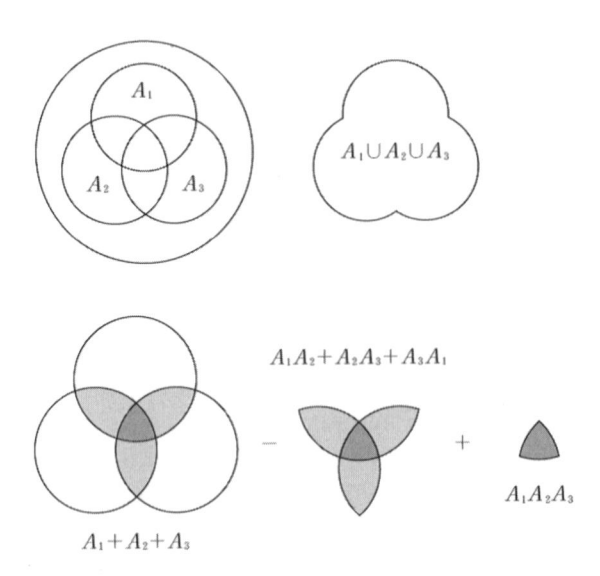

Figure 3: Explanation Using a Venn Diagram ($n = 3$)

Let's look at the relationship with the $3! = 6$ arrangements for the case $n = 3$ shown in Table 3. $A_1 + A_2 + A_3$ combines a total of 6 instances following reseating corresponding to lines 1 and 2, lines 1 and 6, and lines 1 and 3. $A_1A_2 + A_2A_3 + A_3A_1$ combines a total of 3 instances of line 1. $A_1A_2A_3$ corresponds to a single instance of line 1. This yields

$$A_1 \cup A_2 \cup A_3 = 6 - 3 + 1 = 4$$

arrangements in total. There are also two cases when the seatings change, given by lines 4 and 5.

$$1 - A_1 \cup A_2 \cup A_3 = 2$$

There are 4 arrangements in which at least one seating remains unchanged, and 2 arrangements in which they all change. The total of 6 arrangements can thus be confirmed.

4. Montmort's Theory

In this way, for the general case of n students, event A can be obtained by alternating the signs of the terms and summing. The number of arrangements according to which at least 1 student does not change their seat is therefore given by

$$n! - \frac{n!}{2!} + \frac{n!}{3!} - \frac{n!}{4!} + \cdots \pm 1.$$

Since the total number of arrangements is $n!$ the probability P is

$$P = 1 - \frac{1}{2!} + \frac{1}{3!} - \frac{1}{4!} + \cdots \pm \frac{1}{n!}$$

Table 4 shows the probability P versus the number of students n. It is known that as the value of n increases, the value taken by this formula draws closer to

$$1 - \frac{1}{e} \quad (\approx 0.63212).$$

Here e is the natural logarithm with value $e \approx 2.71828$. It's amazing that this probability is basically constant, irrespective of the value of n. Moreover, even for infinitely large n, the value of this formula remains larger than 0.6. Even when reseating a class of 1000 students, or when 1000 people exchange Christmas presents, the chances of a student ending up in the same seat, as well as the chances of someone ending up with the present they brought themselves, are over 0.6. This probability differs from that of shared birthdays, however, in that it does not tend to 1.

n	P
1	1
2	0.5
3	0.6667
4	0.625
5	0.6333
6	0.6320
7	0.6321

Table 4. Number of Students n versus Probability P

According to Feller (1968) this problem has a large number of variations leading all the way back to the brilliant solution by Montmort in 1708 [1]. Two identical sets of n different cards are each arranged in a random order and laid out facing each other. What are the chances of corresponding cards being the same? There are n envelopes and n letters. When a secretary puts the letters in the envelopes indiscriminately, what are the chances of a letter ending up in the right envelope?

The hats deposited in a cloak room are mixed up. Imagine the situation when they are handed back to the guests indiscriminately. When a given person is handed back their own hat, it is considered to be an 'equivalence.' What are the chances of an equivalence occurring? Comparing the chances of an equivalence occurring at an assembly of 8 people and at an assembly of 10000 people, it is surprising to discover that irrespective of the value of n the probability is around $\frac{2}{3}$. The re-seating probability thus has many variations, and continues to receive considerable attention.

Reference

[1] W. Feller, *An Introduction to Probability Theory and Its Applications*, 3rd edition, John Wiley and Sons, (1968).

CHAPTER 21
Opening the Black Box of Random Numbers

Abstract: Random numbers are frequently used all around us, but many people do not know the basis of random numbers. After touching upon prime numbers and prime number theory, this article explains the principles of random number generation using primes and primitive roots, and explains how they are actually handled inside computers.

AMS Subject Classification: 11K45, 00A09, 97A20
Key Words: Random number, Prime number, Sieve of Eratosthenes, Prime number theory, Complex prime number, Primitive root, Congruent linear generator, Mersenne number

1. Random Numbers All around Us

When you leave yourself logged on to a computer for a while, a screen saver is activated. This is to prevent an image from being burned onto the screen. There are a great variety of screen savers, with various different colors and patterns, but one common feature is that the patterns and coordinates are determined using random numbers. The coordinate values are likely generated using a built-in mathematical RAND function. Although many people may know how to use this function, only a few people know how it works. In mathematics, random numbers are related to the prime numbers and primitive roots discussed in elementary number theory, and this time I'd like to think about what's inside the random number black box.

Even without touching a computer, we make great use of random numbers and the concept of random numbers in daily life. Just a few examples include the toss of a coin in soccer matches to determine which team can choose to attack or defend first, the use of dice in *sugoroku* (a Japanese dice game) and *mah jong*, the use of balls with numbers written on them in bingo, and rotating boards and darts used in some lottery games. The thing that can said to be common among all of these is that the outcome is not known in advance. Random numbers are not influenced by previous results, so they can be said to be independent. And since they do not have regularity, they can be said to be irregular.

Dice are often used for random numbers. Dice are regular hexahedrons (cubes) and are capable of generating random numbers between 1 and 6. Dice have dots engraved on them: 6 is opposite 1, 5 is opposite 2, and 4 is opposite 3, so the total is always 7. In order to maintain the center of mass in the exact center of a die, the dot representing 1 is slightly larger than the others. This must be intended to ensure that the numbers from 1 to 6 appear equally on average. Two dice are used for random numbers larger than 6, and other regular polyhedra (such as dodecahedra with 12 faces and icosahedra with 20) are also be used. Books full of random numbers have also been sold in order to avoid having to throw dice. There was indeed a generation when tables of random numbers were good business.

Let's think about the random numbers between 1 and 6 that dice generate. We could consider various random numbers, but we'll restrict ourselves to cycles of 6 here. Let's line up all the numbers from 1 to 6 randomly.

$$1, 2, 3, 4, 5, 6$$

$$6, 5, 4, 3, 2, 1$$

$$1, 3, 5, 2, 4, 6$$

Looking at these, they don't seem like random numbers. Random numbers must be independent; they may not be related to their successors or predecessors. However, when the numbers are lined up like so,

$$2, 6, 4, 5, 1, 3$$

they appear to be random. So why were numbers like these generated? We'll try to answer that here.

2. Primes and Prime Number Theory

Prime numbers are the natural numbers not divisible by any number besides 1 and themselves. For example, $2, 3, 5, \cdots$ and so on are primes.

[Problem 1] Find all the primes between 1 and 100, and count how many there are.

One way to check whether or not the natural number n is prime is to try to divide n by every number from 2 to $n-1$. If none of the numbers cleanly divides it, then it must be prime. Thinking about it carefully, however, it is not necessary to try every number up to $n-1$, and in fact the numbers up to $\sqrt{n}$ are sufficient. Also, by skipping all even numbers, which are divisible by 2, the result can be obtained more efficiently. The problem of finding all the primes up to n is often used as a Visual Basic practical exercise.

A method for finding primes has been known since the age of the ancient Greeks, by using the 'sieve of Eratosthenes' (Figure 1). For example, the way to find the prime numbers up to 30 is as follows. Write out the natural numbers from 2 to 30. Since 2 is prime, fill in the multiples of 2 in a diagonal line, $4, 6, 8, \cdots$ Next, since 3 is prime, fill in the multiples of 3 in a diagonal line, $6, 9, 12, \cdots$ By working in this way, the numbers that remain will be prime. While this method is primitive, it does have the advantage that it prevents oversights. It also wields power when it comes to large computations such as finding all the primes up to 10^7, or counting their number.

	2	3	4̸	5	6̸	7	8̸	9̸	1̸0̸
11	1̸2̸	13	1̸4̸	1̸5̸	1̸6̸	17	1̸8̸	19	2̸0̸
2̸1̸	2̸2̸	23	2̸4̸	25	2̸6̸	2̸7̸	2̸8̸	29	3̸0̸

Figure 1: The Sieve of Eratosthenes

Prime numbers may be applied in various ways. When I used to work for IBM, I knew that prime factorizations have an application in the fast Fourier transform (FFT). When seismic waves are analyzed in terms of frequency, the waves may be converted into sine and cosine Fourier series. If there are 1000 points of sample data, it is necessary to perform 1000 calculations. However, the value of 1024, which is in this neighborhood, has the prime factorization of 2^{10}, and while 1024 calculations would normally be necessary, the FFT, which only requires 10 calculations, may be used. I don't know the details of the algorithm, but I can say that it makes practical use of prime numbers. The FFT was discovered by J.W. Cooley and John Tukey in 1965.

Many aspects of prime numbers have been researched over a long period. One of these is the prime number theorem, which states what proportion of the natural numbers are prime, and was predicted by

Gauss in 1792. It was later proven independently by both Hadamard and Poussin in 1896. It is expressed explicitly by the following formula. Regarding the number of primes between 1 and x, denoted $\pi(x)$,
$$\pi(x) \approx \frac{x}{\log x}$$ ($\log x$ indicates the natural logarithm).
Calculating the actual value for $x < 1000$ reveals that this formula is relatively accurate, but for $x > 10^5$, it is insufficient. A more accurate formula can be obtained using the logarithmic integral,

$$Li(x) = \frac{x}{\log x} + \frac{1!x}{\log^2 x} + \cdots + \frac{(k-1)!x}{\log^k x} + O(\frac{x}{\log^{k+1} x}).$$

The actual number of primes up to the natural number x, the approximate number according to the prime distribution function and the approximate number according to the improved formula are shown in Table 1. $Li(x)$ is taken as the sum up to the 9th term for $x = 10^5$, up to the 10th term for $x = 10^6$, and up to the 12th term for $x = 10^7$. It is thought that this formula is usually accurate for large values of x.

x	Actual number	$\dfrac{x}{\log x}$	$Li(x)$
10	4	4	
100	25	22	
1000	168	145	
10^4	1229	1086	
10^5	9592	8686	9628
10^6	78498	72382	78627
10^7	664579	620421	664915

Table 1. Number of Primes, Actual and Approximated

3. Complex Factors and Prime Numbers

In our discussion of prime numbers, let us touch upon complex factors. Numbers that are only divisible by 1 and themselves are called prime numbers, and we know that $2, 3, 5, \cdots$ are prime. How about expanding numbers in the complex realm? From $2, 3, 5, \cdots$ the numbers 2 and 5 can be factorized as

$$2 = (1+i)(1-i),$$
$$5 = (2+i)(2-i),$$

so 2 and 5 are no longer prime. While this seems a bit strange, this is the truth according to complex numbers. There is such a thing as a prime factorization under complex numbers. $32 + 22i$ is factorized as follows.

$$32 + 22i$$
$$= 2(16 + 11i) = (1 + i)(1 - i)(2 + 3i)(5 - 2i)$$

In this case the complex numbers $1 + i$, $1 - i$, $2 + 3i$ and $5 - 2i$ are the prime factors.

Among the real numbers, the fundamental number is 1. In the case of complex numbers, there are four fundamental numbers, 1, -1, i and $-i$. Numbers which can only be divided by these four factors and themselves are considered prime. By making a computer program using the method of the sieve of Eratosthenes introduced above, it is possible to construct a two-dimensional diagram of the complex numbers that are prime (Figure 2). If you actually make the diagram, you will see that it has a pattern like a tablecloth, and has a rather nice mathematical appearance. See Izumori [1] for information regarding the construction of a suitable BASIC program.

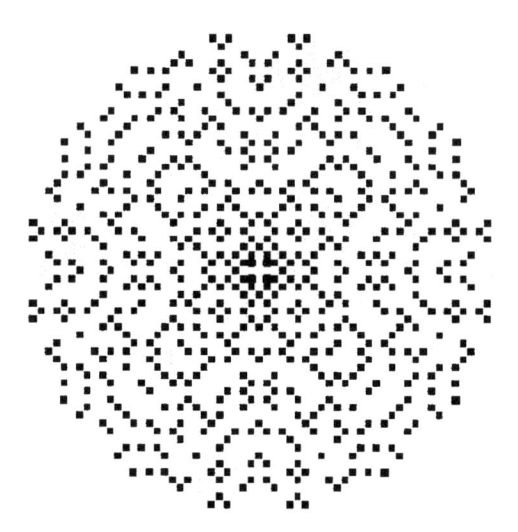

Figure 2: Complex Prime Numbers

4. Primes and Primitive Roots

Leaving the discussion of primes at this point, I'd now like to explain about the pseudo-random numbers that were mentioned at the outset. The numbers 1 to 6 are marked on the faces of dice, and if a sequence like

$$2, 6, 4, 5, 1, 3$$

is generated by some method, they are well suited for use as random numbers. So what is involved in such a method? For primes, there are such things as 'primitive roots'. For example, 3 is a primitive root of the prime number 7. Let's use this relationship to try and generate some pseudo-random numbers.

$3^1 = 3$
$3^2 = 9 \equiv 2 \bmod 7$
$3^3 = 2 \times 3 = 6$
$3^4 = 6 \times 3 = 18 \equiv 4 \bmod 7$
$3^5 = 4 \times 3 = 12 \equiv 5 \bmod 7$
$3^6 = 5 \times 3 = 15 \equiv 1 \bmod 7$

The following sequence of random numbers is thus generated in this way.

$$3, 2, 6, 4, 5, 1$$

The manner of calculation is repeated multiplication by 3. When the result is greater than 7, a multiple of 7 is subtracted and the remainder is taken as the answer. After repeating this calculation 6 times, the result returns to 1.

This is called a linear congruential generator. In this case, the last number is 1, so to avoid this, if the initial value is taken as $3^2 \equiv 2 \bmod 7$ then the sequence becomes

$$2, 6, 4, 5, 1, 3.$$

The random numbers from 1 to 6 can be generated from the prime number 7. Putting it another way, the numbers $3^1, 3^2, 3^3, 3^4, 3^5, 3^6$ with 7 as a modulus can be said to constitute a cyclic group.

For a prime number p and coprime number r, the relationship $r^{p-1} \equiv 1 \bmod p$ is satisfied. This is known as Fermat's little theorem. The reason that the theorem is described as 'little' is to distinguish it from the famous Fermat's last theorem. This congruence expression may have solutions of the form $r^i \equiv 1 \bmod p$ for $i < p-1$, but r is only a primitive root of p when the expression is satisfied for $i = p-1$. For the prime number 7, besides 3, 5 is also a primitive root. The primitive roots of the primes 3, 5, 7, 11 and 13 are shown in Table 2.

[Problem 2] Using the fact that 6 is a primitive root of the prime number 11, generate a pseudo-random sequence from 1 to 10.

p (Prime)	r (Primitive Root)
3	2
5	2, 3
7	3, 5
11	2, 6, 7, 8
13	2, 6, 7, 11

Table 2. Primes and Primitive Roots

5. Random Numbers on Computers

The explanations above deal with the principles of random number generation, but how are random numbers actually handled inside computers? Computers have built-in functions such as RAND for generating random numbers in the interval $(0, 1)$. Integer random numbers are generated first, and then divided by the largest possible number to calculate a random real number in the interval $(0, 1)$. I don't know the details of the algorithm used in the Visual BASIC built in RAND function, but one well-known pseudo-random number generation algorithm is as follows. All such random number algorithms involve the concept represented by the following congruence equation.

$X_i = (aX_{i-1} + c) \bmod M$ (a and c are constants)

After setting X_{i-1} to some value, it is multiplied by a. If this value exceeds M, it is divided by M and the remainder is given as the value of X_i. Random numbers are generated one after the other using this iterative formula.

The variables handled by programs include real number types and integer types. Integer type random numbers use 4 bytes (*i.e.*, 32 bits). The number of different values that can be represented with 32 bits is a power of 2, but one bit at the head is the 'sign bit.' A 0 signifies a positive number, while a 1 signifies a negative number. The actual values are expressed using the 31 bits that remain after excluding this one sign bit. Assuming that the random numbers are positive, the largest number that can be represented by 31 bits is $2^{31} - 1 = 2147483647$.

By generating random numbers in the interval 1 to $2^{31} - 1$ using the congruence expression explained above and dividing them by this largest number, random numbers can be generated in the interval $(0, 1)$.

The problem is then how to select the prime p and the primitive root r. The value of the largest integer-type variable is $2^{31} - 1$. The primes that can be expressed in the form $2^p - 1$ are known as Mersenne numbers. There are 23 Mersenne numbers existing in the range $p < 11400$, including

$$p = 2, 3, 5, 7, 13, 17, 19, 31, 61, \cdots$$

These values of p do not include all the primes.

Fortunately, the largest number that can be expressed with 31 bits, $2^{31} - 1$, is a Mersenne number and is prime. It is therefore useful to obtain the primitive roots of $2^{31} - 1$. As shown in Table 2, the number of primitive roots is not just 1. For small prime numbers, the primitive roots may be obtained easily, but for a number like $2^{31} - 1$ which has 10 digits in base 10, it's not so easy to obtain them and we must rely on the help of a computer.

S.K. Park and K.W. Miller used $M = 2^{31} - 1$ as a prime, and $a = 7^5$ as a primitive root in the formula

$$X_i = 7^5 X_{i-1} \bmod (2^{31} - 1).$$

As mentioned in reference [2], finding suitable pairs of primes and primitive roots is difficult. $2^{31} - 1 = 2147483647$ is prime, but the primitive root $7^5 = 16807$ is not prime. The numbers $M = 2^{31} - 1$ and $a = 7^5$ used in the congruence formula are coprime, so the above stated Fermat's little theorem is satisfied. That is,

$$(7^5)^{2^{31}-2} \equiv 1 \bmod (2^{31} - 1).$$

This is satisfied without the need for calculation, but actually verifying it with a computer confirms that after $2^{31} - 2$ iterations, the value returns to 1 for the first time. Because the result does not return to 1 at any point before $2^{31} - 2$ iterations, it passes through every possible value and is thus an excellent primitive root.

The method above has a cycle length of $2^{31} - 2$ which prompts no complaints, but its weak point is that the division calculation required by the $\bmod(2^{31} - 1)$ operation takes time. This issue is addressed by IBM's well-known improved RANDU subroutine (1970). The RANDU congruence expression is given by the following formula.

$$X_i = 65539 X_{i-1} \bmod 2^{31}$$

$M = 2^{31}$ ($= 2147483648$) is not prime, so Fermat's little theorem does not apply, but $a = 2^{16} + 3$ ($= 65539$) is prime, and M is coprime with a, so the generation of random numbers with a large cycle length can be expected. Confirming this with a computer revealed that

$$(2^{16} + 3)^{2^{29}} \equiv 1 \bmod 2^{31}.$$

The cycle length is a quarter of the maximum cycle length, which is a relatively long cycle.

From a mathematical perspective it is beneficial to choose a prime for M. In the case of RANDU, however, the divisor 2^{31} is not prime. The reason it was chosen is surely because it yields an efficient division calculation. Inside a computer, a division by 2^{31} is performed using a machine instruction that slides a binary number 31 bits to the right. This uses a shift register. Leaving the details for another time, suffice it to say that this method has the advantage of being an especially fast computation. But RANDU belongs to the year 1970. As computer technology develops, the speed of computation increases and this kind of ingenuity is no longer necessary. Due to the lack of precision in the random numbers generated by RANDU, other methods are now in use.

Random numbers are generated using primes and primitive roots. Perhaps by knowing this mechanism, one may understand and appreciate the wonder of mathematics.

References

[1] H. Izumori, *PC de Tanoshimu Koko Sugaku* [*Enjoying High School Maths with a Computer*], Tokyo: Saiensusya, 47-51, (1991).

[2] S.K. Park, K.W. Miller, Random number generators: Good ones are hard to find, *Communication of the ACM*, 31, 1192-1201, (1988).

CHAPTER 22
Calculating $\sqrt{2}$

Abstract: This article concerns root 2. The ratio of height and width of copy paper is root 2, and can be calculated by using similar diagrams. Root 2 is an irrational number whose cardinality is uncountably infinite. Roots are calculated numerically by the Newton-Raphson method which uses an approximating 2nd order function and its tangent.

AMS Subject Classification: 68W02, 00A09, 97A20

Key Words: Ratio of height and width of copy paper, Similar diagrams, Irrational number, *Reductio ad absurdum*, Cardinality of numbers, Cantor's diagonal method, Countably infinite and uncountably infinite, 2nd order function and its tangent, Newton-Raphson method

1. The Height and Width of Photocopier Paper

I wondered exactly how much students in the humanities know about root 2, so I took this up in a lecture on information mathematics. Root 2 is written $\sqrt{2}$, and produces 2 when it is squared. I had my students discuss this value themselves to see what they would say. Students who didn't know its value already made the following kinds of calculation.

$1^2 = 1$ and $2^2 = 4$ so $1 < \sqrt{2} < 2$

$1.5^2 = 2.25$ and $1.4^2 = 1.96$ so $1.4 < \sqrt{2} < 1.5$

They explained how to obtain the value of root 2 within a certain level of accuracy by means of written calculations using this bisective method. Some students knew the values of root 2 and root 3 without the need for calculation according to mnemonics like "I wish I knew - the root of two. O charmed was he - to know root three. So we now strive - to find root five" (which encodes the values 1.414, 1.732 and 2.235).

A4 and B4 copy paper sizes are of a familiar size, and the ratio of their height and width is 1 to root 2, although surprisingly few people know this. This ought to have been learned in junior high-school geometry classes or in the first year of high school, but whether mathematics was not necessary for their examinations, or whether they just hated mathematics, many students forget this fact. I therefore give students the following problem to think about.

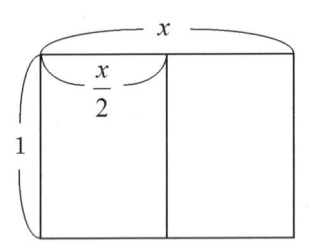

[Problem 1] Find the ratio of the height and width of a piece of copy paper.

Students reply with '1 to 2' or '2 to 3' according to the appearance of the paper. At that point I demonstrate that copy paper has the amazing property that folding it in half does not change the ratio of the height and width. Such shapes are known as similar diagrams. Maybe students are learning this for the first time - they always seem newly impressed.

The ratio of the height and width may be obtained using this property of similarity. Let's denote the ratio of the height and width of a rectangle as x to 1. Thinking about the same rectangle folded in half reveals that the height is $\frac{x}{2}$ while the width is 1, so

$$1 : x = \frac{x}{2} : 1.$$

Since the product of the inner terms is equivalent to the product of the outer terms (this should have been learned at junior high-school),

$$x \times \frac{x}{2} = 1 \times 1.$$

Solving this yields $x = \sqrt{2}$. That is, the ratio of the height and width of copy paper is $1 : \sqrt{2}$.

Figure 1: The Ratio of the Height and Width of Copy Paper

2. Proving That Root 2 Is an Irrational Number

It is known that root 2 is an irrational number. In response to this information, students ask what irrational numbers are, and I reply that they are numbers which continue forever after the decimal point. "Doesn't that mean that numbers like $0.999999\cdots$, which go on forever are irrational?" they ask.

so, $\dfrac{1}{3} = 0.333333\cdots \qquad \dfrac{1}{3} \times 3 = 0.999999\cdots$

$1 = 0.999999\cdots$

This is not an irrational number. To put it precisely, irrational numbers are those numbers which continue indefinitely after the decimal point, without cycling. Numbers like

$\dfrac{2}{15} = 0.1333333\cdots$, and $\dfrac{1}{7} = 0.1428571428571\cdots$

and so on continue indefinitely, but when the part immediately after the decimal point cycles, the number is not known as irrational. Recurring decimals can be expressed as vulgar fractions by calculating their value as an infinite series.

Rational numbers can be expressed as a ratio of integers, but irrational numbers cannot be expressed as an integer ratio. Few students can reply correctly up to this level. In Japanese, 'irrational numbers' are known as *murisu*, but the part corresponding to 'irrational' also has other meanings, and the name is not so appropriate. During the *Meiji* era, the English terms 'rational number' and 'irrational number' were translated into Japanese as *yurisu* and *murisu*, but the real meaning is whether or not they can be expressed as a ratio of integers, so some mathematical historians have suggested that the terms *yuhisu* and *muhisu*, meaning, respectively, 'number with a fraction' and 'number without a fraction,' are more appropriate.

[Problem 2] Prove that $\sqrt{2}$ is an irrational number.

Well then, let's try and prove that root 2 is irrational, *i.e.*, that it cannot be expressed as a ratio of integers. This is explained in the high-school mathematics syllabus in Japan, but there is little chance of the proof appearing in university entrance examinations, so many students avoid studying it and do not know the method. It is important, so I'd like to reiterate the proof here. The proof utilizes *reductio ad absurdum*.

Suppose that root 2 could be expressed as a ratio of integers as follows.

$\sqrt{2} = \dfrac{q}{p}$ (where p and q are mutually prime natural numbers)

The fact that p and q are mutually prime expresses a requirement that they constitute an irreducible fraction. Squaring both sides of this equation and eliminating the denominator yields

$$q^2 = 2p^2.$$

From this equation, q^2 is a factor of 2. If q^2 is a factor of 2, then q is also a factor of 2 (this part of the proof is a little tricky). We can thus write,

$q = 2m$ (where m is a natural number).

Substituting this into the previous equation and rearranging yields the following.

$$4m^2 = 2p^2$$

$$p^2 = 2m^2$$

From this equation p^2 is a factor of 2. Again, if p^2 is a factor of 2, then p is also a factor of 2. Putting these results together, q is a factor of 2 and p is a factor of 2. This contradicts the assumption that p and q are mutually prime. Thus $\sqrt{2}$ cannot be expressed as a fraction $\frac{q}{p}$ (an integer ratio), and is therefore an irrational number.

3. The Cardinality of Rational Numbers and Irrational Numbers

How many rational numbers and how many irrational numbers are there? Which type describes the most numbers? At this point allow me to introduce a rather interesting topic. It should have been learned that according to Euclidean geometry "the sum of two edges of a triangle is larger than the other edge, and the difference between two edges is less than the other edge." However, let's try to prove that "the sum of two edges of a triangle is equal to the other edge" or alternatively "all the line segments have equal length!" Of course this is sophistry, but can we identify where the mistake lies?

Suppose we have the triangle ABC shown in Figure 2. Draw B'C' parallel to the base edge BC. In general "lines are a collection of points," "faces are a collection of straight lines" and "solids are a collection of faces." The line segment BC is packed with points. As a representative example, let us select the point P. Connecting this point P and the vertex A produces a line that crosses B'C', and the intersection point is denoted Q. Every point on line segment BC has a corresponding point on line segment B'C'. We postulated that a line segment was a collection of points, and thus the line segments BC and B'C' are equivalent.

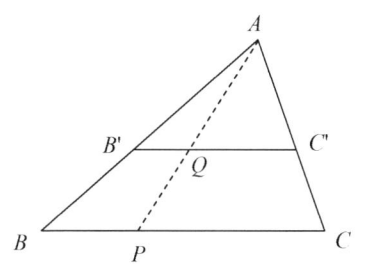

Figure 2: Are Line Segments BC and B'C' Equivalent?

Around the time I was a first-year university student there was an aspect of mathematics that I had just learned, and all I wanted to do was talk about it with my friends. The particular piece of sophistry in question identified a mistake in the notion that "lines are a collection of points." No matter how many points are gathered, they do not constitute a line. They are just a collection of points. This is just how difficult it is to comprehend infinity. In this regard, there are an infinite number of natural numbers, rational numbers and irrational numbers, but the natural numbers and rational numbers form infinite sets that may be enumerated (they are 'countable'), while the irrational numbers form an infinite set that cannot be enumerated (they are 'uncountable').

In 1891, Cantor proved that there is no 1-1 mapping from the natural numbers to the real numbers in the interval $(0, 1]$ using the so-called diagonal method. This is because the cardinality of any set which has a 1-1 mapping with the set of natural numbers must be equal to the cardinality of the natural numbers. This cardinality is written $\aleph_0$, and read 'aleph zero.' There is a 1-1 correspondence between the rational numbers and the natural numbers, so the cardinality of the rational numbers is $\aleph_0$. $\aleph$ is a Hebrew letter corresponding to the letter A. Sets with a cardinality of $\aleph_0$ are known as enumerable, or 'countably infinite.' In contrast, the cardinality of the real numbers is written $\aleph$, and $\aleph_0 < \aleph$.

4. Tangential Line Equations

The explanation is now for the most part complete. The theme in this chapter was the calculation of root 2. If the 2nd order function and the corresponding tangent are known, then root 2 can be obtained efficiently. The equation for the tangent can be found using the ideas of differentiation learned in high-school mathematics in Japan. However, students in the humanities for whom differentiation and integration were not necessary for university entrance examinations may complain of not having learned these topics in school classes. In such a case a compromise may be reached with students by explaining a little advance knowledge regarding the 2nd order functions dealt with in Mathematics I.

2nd order functions are functions which are proportional to a 2nd order term. Students at junior high school learn about 1st order functions which are straight lines proportional in the 1st order, but in high school they learn about 2nd order curved-line functions which are proportional in the 2nd order. Having students draw diagrams of such 2nd order functions on note paper can be used to confirm that this has been learned.

I then put forward the problem of obtaining the equation for the tangent which touches such a 2nd order function.

[Problem 3] Find the equation of the tangent which touches $y = x^2$ at $x = 2$.

Many students complain that not having learned differentiation, they cannot use derivative-based methods, so I explain a method of solving the problem using only Mathematics I knowledge without relying on differentiation. The 2nd order function is written $y = x^2$, and its tangent is written $y = ax + b$. The fact that they touch means that the 2nd order function and the straight line "intersect at a point" (Figure 3). Let us therefore seek the solution to the two equations according to the condition that it is a repeated solution (*i.e.*, there is only one solution).
$$x^2 = ax + b$$
$$x^2 - ax - b = 0$$
From the discriminant, $D = a^2 + 4b = 0$,
$$b = -\frac{a^2}{4}.$$
The tangent passes through the point $P(x, x^2)$ on the 2nd order function, so substituting this into the equation for the tangent yields
$$x^2 = ax - \frac{a^2}{4}.$$
Solving this yields
$$4x^2 = 4ax - a^2,$$
$$a^2 - 4xa + 4x^2 = 0,$$
$$(a - 2x)^2 = 0.$$
The constants in the equation for the tangent may thus be determined.
$$a = 2x$$

$$b = -\frac{4x^2}{4} = -x^2$$

Compiling the results above, the equation for the tangent which intersects the function at $P(x, x^2)$ may be expressed in terms of X and Y as follows.

$$Y = 2xX - x^2$$

Substituting $x = 2$ into this equation yields

$$Y = 4X - 4.$$

The equation of the tangent line is thus

$$y = 4x - 4.$$

This calculation can certainly be made without using differentiation.

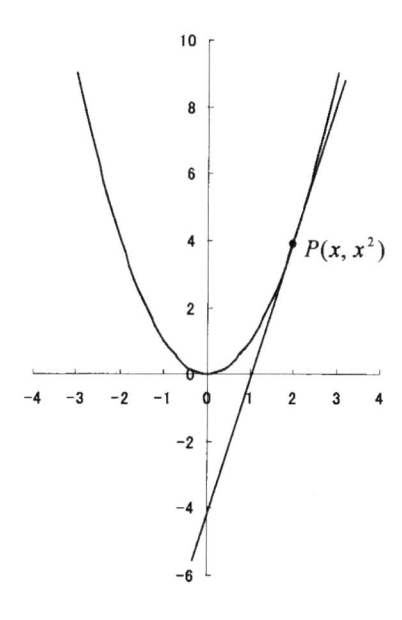

Figure 3: 2nd Order Function and Its Tangent

5. The Newton-Raphson Method

The equation of the tangent was calculated above using Mathematics level I knowledge, so now let's try using differentiation to find the tangent. Firstly, let's try to find the gradient of the tangent to the 2nd order function $y = x^2$ at the point $P(x, y)$. The gradient of the tangent may be found as the differential coefficient. The idea behind the differential coefficient is as follows. To begin with, it does not involve thinking of $P(x, y)$ as having an intersection at a single point, but rather as intersecting two points such that the second has its x coordinate offset by a small distance h, $Q(x + h, (x + h)^2)$. The slope between P and Q may then be calculated.

$$\frac{\Delta y}{\Delta x} = \frac{(x+h)^2 - x^2}{(x+h) - x} = \frac{2hx + h^2}{h} = 2x + h$$

Since the intersection does in fact occur at a single point, Q is brought close to P. That is to say, Δx is brought close to 0. When Δx tends to 0, Δy also approaches 0. You might think that since both the numerator and the denominator are close to 0, the value of the fraction could not be obtained, but it may in fact be properly calculated. This kind of manipulation is known as a limit. This value is the gradient of the tangent, and the derivative function is expressed as $f'(x)$.

$$f'(x) = \lim_{\Delta x \to 0} \frac{\Delta y}{\Delta x} = \lim_{h \to 0} \frac{2hx + h^2}{h} = 2x$$

Also, since the tangent passes through the point $P(x, x^2)$, expressing the equation of the tangent in terms of the coordinates (X, Y) yields

$$Y - x^2 = 2x(X - x)$$
$$Y = 2xX - x^2.$$

This is the same as that obtained using the discriminant method.

Well then, the focus of this chapter was the calculation of root 2, and in the case of root 2, the 2nd order function

$$y = x^2 - 2$$

is used. This is just the original 2nd order function shifted along the Y-axis by -2. The equation of the tangent to this 2nd order function can be obtained. The calculation is easy because the derivative function can be used. In general, the equation of the tangent is

$$Y - f(x) = f'(x)(X - x),$$

and substituting $f(x) = x^2 - 2$ and $f'(x) = 2x$ into this equation yields

$$Y - (x^2 - 2) = 2x(X - x).$$

Rearranging this,

$$Y = 2xX - x^2 - 2.$$

Now then, let's use this formula to find the value when the tangent cuts the X-axis, i.e., when $Y = 0$.

$$X = \frac{x}{2} + \frac{1}{x} = \frac{1}{2}(x + \frac{2}{x})$$

The value $x = x_1$ may be used temporarily in the right-hand side of this equation. Substituting this into the right-hand side and representing the value calculated as $X = x_2$ yields a recurring equation with the following form.

$$x_2 = \frac{1}{2}(x_1 + \frac{2}{x_1})$$

The initial value of x_1 may be set, for example, as $x_1 = 1$. This value is substituted into the equation above and the value of x_2 is calculated. Then for the next candidate for the value of x_1, the calculated value of x_2 replaces the original value of x_1. If this process is repeated a number of times, it is possible to calculate the value of root 2 by hand. Table 1 shows that the result obtained after 3 iterations of this calculation is 1.414215, which is correct to 5 decimal places. This method is known as the Newton-Raphson method, taking the name of the man who thought up the derivative method, Newton.

When roots are calculated in BASIC or C programs, it is sufficient to set the error tolerance at which to break off iterating the calculations so as to stopping when the absolute value of the difference between x_1 and x_2 falls below a certain value ϵ (e.g., 10^{-7}).

Number of iterations	x_1	x_2
1	1	$3/2 = 1.5$
2	$3/2$	$17/12 = 1.416$
3	$17/12$	$577/408 = 1.414215$

Table 1. Calculation of Root 2 Using a Recurring Formula.

[Problem 4] Obtain the value of root 5 to within 3 decimal places using the methods introduced for root 2.

Computers have built-in functions like SQRT, SQR, etc. for calculating roots. Calculators have a button with the root symbol ($\sqrt{\ }$) marked on it. Both computers and calculators perform iterative calculations according to the Newton-Raphson method, and do not contain root data prepared in advance and stored as a database.

The capabilities of personal computers and calculators have really come along. Between the 1960s and early 1970s we didn't have personal computers, calculators were very expensive and the technology was in the process of development. There was a difference in the time taken by the four arithmetic operations and the calculation of roots which could be confirmed with the naked eye. The result of an addition or subtraction appeared in an instant when the Enter key was pressed, but a root calculation took a few seconds. There were roots for which the result was revealed immediately, and roots which took time to calculate. Those were the good old days when you could stare at a busy calculator and think about the effort it was putting into the iterative calculations we have been discussing in this chapter.

CHAPTER 23
The Brachistochrone Curve: The Problem of Quickest Descent

Abstract: This article presents the problem of quickest descent, or the Brachistochrone curve, that may be solved by the calculus of variations and the Euler-Lagrange equation. The cycloid is the quickest curve and also has the property of isochronism by which Huygens improved on Galileo's pendulum.

AMS Subject Classification: 34A02, 00A09, 97A20
Key Words: Brachistochrone curve, Law of energy conservation, Calculus of variations, Euler-Lagrange equation, Cycloid, Isochronism, Huygens's pendulum

1 Which Is the Quickest Path?

Suppose there is an incline such as that shown in Figure 1. When a ball rolls from A to B, which curve yields the shortest duration? Let's assume that we have three hypotheses: a straight line, a quadratic and a cycloid. The shortest path from A to B is the straight line, so one might think that the straight path is the fastest, but in fact it is surprisingly slow. It's better to select a path which has a downward drop in order to accelerate the ball in the first phase, so that it rolls quickly. The ball arrives earlier on the quadratic path than on the straight line path. However, increasing the degree of the function causes the ball to travel more slowly on the flat section.

It is said that Galileo (1564-1642) first presented this problem. It is also known that the cycloid is the curve which yields the quickest descent. This time I will discuss this problem, which may be handled under the field known as the calculus of variations, or variational calculus in physics, and introduce the charming nature of cycloid curves.

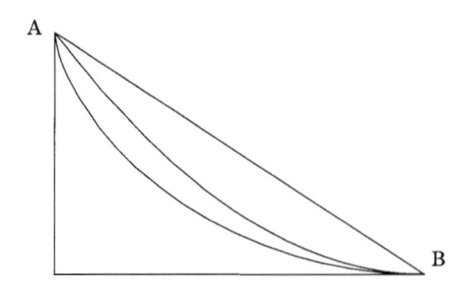

Figure 1: Which Path Yields the Shortest Duration?

2 Model Construction and Numerical Computation

Before obtaining the form of the curve analytically, let's try some numerical calculations in order to gain a rough understanding of the problem. I calculated the arrival time for several different curves on a computer using some spreadsheet software. Using the coordinate frame shown in Figure 2, the ball was assumed to roll from a point at a height of y_0. The only force acting on the ball is the force due to gravity, mg, and from the law of energy conservation, the sum of the potential energy and the kinetic energy is constant, so when the ball is at a height of y, the speed v may be obtained as follows.

$$mgy_0 = \frac{1}{2}mv^2 + mgy \qquad (2.1)$$

$$v = \sqrt{2g(y_0 - y)} \qquad (2.2)$$

When the shape of the curve is fixed, the infinitesimal distance ds may be found, and dividing this by the velocity v yields the infinitesimal duration dt. If an infinitesimal duration such as this dt is integrated, the result is the time until arrival.

$$ds = \sqrt{dx^2 + dy^2} \qquad (2.3)$$

$$dt = \frac{ds}{v} \qquad (2.4)$$

$$T = \int dt \qquad (2.5)$$

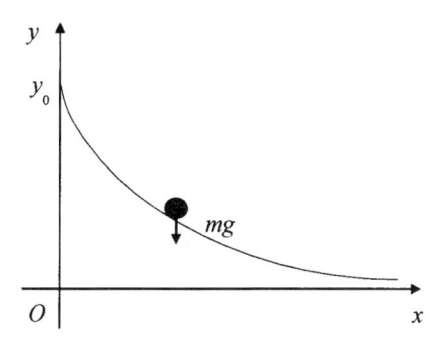

Figure 2: Gravity Is the Only External Force

A piecewise curve with 100 divisions, a height of 2 m and a width of $\pi (= 3.14)$ m was used to produce numerical data. The results revealed an arrival time of 1.189 s for the straight line, 1.046 s for the quadratic, 1.019 s for a cubic curve, 1.007 s for an ellipse and 1.003 s for the cycloid. The straight line was the slowest and the curved line the quickest. The difference between the ellipse and the cycloid was slight, being only 0.004 s.

The arrival times were confirmed with a computer, but this lacks a sense of reality, which made me want to build an actual model. I wanted to make a large model, but considering the cost of construction and storage I considered a cut-down model. I found some plywood with horizontal and vertical dimensions of 30 × 45 cm in a DIY store. The block was 1.2 cm thick, and since *pachinko* balls (used in the popular Japanese version of pinball) are 1.1 cm in diameter, this was sufficient. On the computer, the time for the straight line was 0.445 s, for the quadratic it was 0.391 s and for the cycloid it was 0.375 s. Performing these experiments in reality, the difference between the cycloid and the straight line was clear, but the difference between the cycloid and the quadratic required appropriate caution. The difference in arrival times corresponded to those for a single *pachinko* ball.

3 The Calculus of Variations and Functional Integrals

The numerical calculations above were made after the form of the curve was known, but let's think about the problem of minimizing the arrival time in the case that the form of the curve is not known. Let us denote the starting and final locations by A and B, respectively [1]. The integral from the time-step when the ball is at the starting location, t_A, and the time-step of arrival, t_B, is the duration of motion, T.

$$T = \int_{t_A}^{t_B} dt \tag{3.1}$$

Let us investigate the expression of this dt using x, y and y'. If the infinitesimal element is taken as ds, then the following relationship may be established by Pythagoras' theorem.

$$ds = \sqrt{dx^2 + dy^2} = \sqrt{1 + \left(\frac{dy}{dx}\right)^2} \, dx \tag{3.2}$$

The speed of the ball, v, may be found by taking the time derivative of the distance along the curve. This may be written as follows.

$$v = \frac{ds}{dt} = \frac{ds}{dx}\frac{dx}{dt} = \sqrt{1 + \left(\frac{dy}{dx}\right)^2} \frac{dx}{dt} \tag{3.3}$$

Equation (3.1) may be rewritten as follows, using Equations (3.2) and (3.3).

$$T = \int_{t_A}^{t_B} dt = \int_{x_A}^{x_B} \frac{dt}{ds}\frac{ds}{dx} dx = \int_0^x \frac{\sqrt{1 + \left(\frac{dy}{dx}\right)^2}}{v} dx \tag{3.4}$$

If the y coordinate is taken as being in the downwards direction, then the distance fallen, y, and the speed, v, must obey the principle of energy conservation so the equation

$$\frac{1}{2}mv^2 = mgy \tag{3.5}$$

is satisfied. Rearranging yields,

$$v = \sqrt{2gy}. \tag{3.6}$$

Substituting this into Equation (3.4), and writing y' for $\frac{dy}{dx}$ yields the following equation.

$$T = \int_0^x \sqrt{\frac{1 + \left(\frac{dy}{dx}\right)^2}{2gy}}\, dx = \int_0^x \sqrt{\frac{1 + y'^2}{2gy}}\, dx \qquad (3.7)$$

How should one go about minimizing T according to this equation? The selection of the integrand

$$L(x, y, y') = \sqrt{\frac{1 + y'^2}{2gy}} \qquad (3.8)$$

in order to minimize T is a problem in the calculus of variations.

4 The Euler-Lagrange Equation

Now, given a function $L(x, y, y')$, let's think about the problem of finding the extremal value of the integral,

$$I = \int_{x_1}^{x_2} L(x, y, y')\, dx, \qquad (4.1)$$

by setting the function $y(x)$. I is called a functional. This expresses the meaning that, in comparison to a normal function, it is a "function of a function." Suppose we have a function which is slightly offset from $y(x)$, the function we are seeking is

$$Y(x) = y(x) + \epsilon \delta(x). \qquad (4.2)$$

Its integral is

$$I(\epsilon) = \int_{x_1}^{x_2} L(x, Y, Y')\, dx. \qquad (4.3)$$

Consider the condition according to which it takes its extreme value,

$$\left.\frac{dI(\epsilon)}{d\epsilon}\right|_{\epsilon=0} = 0. \qquad (4.4)$$

Y and Y' both depend on ϵ. Bearing in mind that $\dfrac{dY}{d\epsilon} = \delta(x)$ and $\dfrac{dY'}{d\epsilon} = \delta'(x)$, and taking the derivative yields

$$\left.\frac{dI(\epsilon)}{d\epsilon}\right|_{\epsilon=0} = \int_{x_1}^{x_2} \left(\frac{\partial L}{\partial Y}\delta(x) + \frac{\partial L}{\partial Y'}\delta'(x)\right)\bigg|_{\epsilon=0} dx. \qquad (4.5)$$

Integrating the second term on the right-hand side by parts yields

$$\int_{x_1}^{x_2} \frac{\partial L}{\partial Y'}\delta'(x)dx = \frac{\partial L}{\partial Y'}\delta(x)\Big|_{x_1}^{x_2} - \int_{x_1}^{x_2} \frac{d}{dx}\left(\frac{\partial L}{\partial Y'}\right)\delta(x)dx. \quad (4.6)$$

For x_1 and x_2, $\delta(x) = 0$ so the first term on the right-hand side of this equation is zero. The condition for the extremal value thus becomes

$$\frac{dI(\epsilon)}{d\epsilon}\Big|_{\epsilon=0} = \int_{x_1}^{x_2}\left(\frac{\partial L}{\partial y} - \frac{d}{dx}\frac{\partial L}{\partial y'}\right)\delta(x)dx = 0. \quad (4.7)$$

Since $\delta(x)$ is arbitrary, the following condition determining $y(x)$ may be obtained.

$$\frac{\partial L}{\partial y} - \frac{d}{dx}\left(\frac{\partial L}{\partial y'}\right) = 0 \quad (4.8)$$

This formula is known as Euler's equation, or alternatively the Euler-Lagrange equation. The calculus of variations originates in Fermat's principle which expresses how the path of a beam of light varies as it passes through media with different refractive indices. This operates according to the principle that the path is selected in order to minimize the passage time.

5 Solving Euler's Equation

Now then, $L = \sqrt{\dfrac{1+y'^2}{2gy}}$ may be substituted into the Euler-Lagrange equation $\dfrac{\partial L}{\partial y} - \dfrac{d}{dx}\left(\dfrac{\partial L}{\partial y'}\right) = 0$, but x is not explicitly contained in L, i.e., it is a function of y and y' alone. The following transformation of Euler's equation may therefore be used.

$$L - y'\left(\frac{\partial L}{\partial y'}\right) = C \quad (5.1)$$

Substituting for L in this equation,

$$\sqrt{\frac{1+y'^2}{2gy}} - y'\left(\frac{y'}{\sqrt{2gy(1+y'^2)}}\right) = \frac{1}{\sqrt{2gy(1+y'^2)}} = C. \quad (5.2)$$

Squaring both sides, the equation may be rearranged as follows. Since the right-hand side is constant, so we may write it as $2A$.

$$y(1 + y'^2) = \frac{1}{2gC^2} = 2A \tag{5.3}$$

Equation (5.3) is rearranged as follows.

$$y' = \sqrt{\frac{2A - y}{y}} \tag{5.4}$$

The domain of the curve is taken as

$$2A \geq y \geq 0. \tag{5.5}$$

The initial condition is taken as $y = 0$ when $\theta = 0$. At this point y may be rewritten as the following parametrical expression using a change of variable.

$$y = A - A\cos\theta \tag{5.6}$$

This change of variable (5.6) may seem somewhat sudden. Rather than determining the nature of the function according to the calculus of variations, in this case it was already known that the cycloid is the curve of quickest descent because research on cycloids has been developing for a considerable length of time. It is sufficient to understand that this curve was taken as a hypothesis and the solution was obtained using the calculus of variations.

If both sides of Equation (5.6) are differentiated then the result is as follows.

$$dy = A\sin\theta\, d\theta = 2A\cos\frac{\theta}{2}\sin\frac{\theta}{2}\, d\theta \tag{5.7}$$

It is possible rewrite Equation (5.7) using the parametric expression for y as follows.

$$y' = \sqrt{\frac{2A - y}{y}} = \sqrt{\frac{A + A\cos\theta}{A - A\cos\theta}} = \sqrt{\frac{\cos^2\frac{\theta}{2}}{\sin^2\frac{\theta}{2}}} = \frac{\cos\frac{\theta}{2}}{\sin\frac{\theta}{2}} \tag{5.8}$$

This can be written as follows, by multiplying both sides by dx.

$$dy = -\frac{\cos\frac{\theta}{2}}{\sin\frac{\theta}{2}} dx \tag{5.9}$$

The relationship between x and θ may be found by collating Equations (5.7) and (5.9), and eliminating dy.

$$dx = 2A\sin^2\frac{\theta}{2}d\theta = A(1-\cos\theta)d\theta \tag{5.10}$$

Integrating both sides,

$$x = A(\theta - \sin\theta) + D. \tag{5.11}$$

For the initial condition $x = 0$ when $\theta = 0$, the constant of integration is $D = 0$. Finally, the parametric expression for the curve of quickest descent is as follows.

$$x = A(\theta - \sin\theta) \tag{5.12}$$

$$y = A(1 - \cos\theta) \tag{5.13}$$

6 The Cycloid

A cycloid may be expressed as the trajectory of a point fixed on the circumference of a circle with radius a when the circle rolls along a straight line. When the angle of the circle's rotation is θ, the coordinates on the curve are as follows.

$$x = a(\theta - \sin\theta) \tag{6.1}$$

$$y = a(1 - \cos\theta) \tag{6.2}$$

The derivatives are $\frac{dx}{d\theta} = a(1 - \cos\theta)$ and $\frac{dy}{d\theta} = a\sin\theta$. Bearing in mind that $\cos\theta = 1 - \frac{y}{a}$, we may write

$$\left(\frac{dy}{dx}\right)^2 = \frac{(dy/d\theta)^2}{(dx/d\theta)^2} = \frac{2a - y}{y}. \tag{6.3}$$

This is the differential equation of the cycloid, and it should be noted that it is equivalent to the previously stated Equation (5.4). It is common for the explanation of the cycloid given in high-school mathematics textbooks to state no more than that it is the trajectory of a point on a bicycle wheel. It's a shame that the exceptional property that it is the curve of quickest descent is rarely explained.

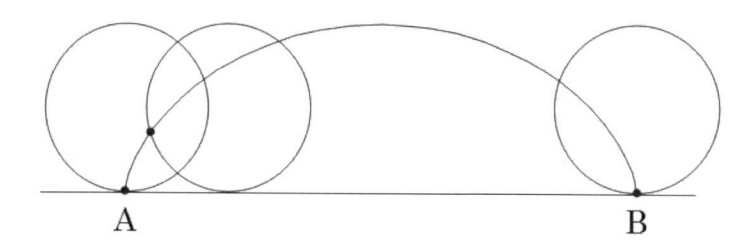

Figure 3: The Cycloid

7 Isochronism

It was Galileo who discovered the isochrosism of pendulums. Supposing the length of a given pendulum is l, the gravitational constant is g and the equilibrium point of the pendulum is at the origin, then when the swing angle of the pendulum is θ, the equation of the pendulum's motion is as follows.

$$ml\frac{d^2\theta}{dt^2} = -mg\sin\theta \tag{7.1}$$

Hypothesizing that when the angle of swing is small, $\sin\theta \approx \theta$, the equation may be simplified as follows.

$$l\frac{d^2\theta}{dt^2} = -g\theta \tag{7.2}$$

The period is

$$T = 2\pi\sqrt{\frac{l}{g}}. \tag{7.3}$$

I explained that the cycloid was the curve of quickest descent, but it has one more exceptional property, it is isochronic. Whether a ball is rolled from the point A shown in Figure 4, or from the intermediate point C, the time taken to arrive at point B is the same.

CHAPTER 23

[Question 1] Prove that for balls placed on a cycloid curve, even if their positions differ, the time taken to reach the lowest point is the same.

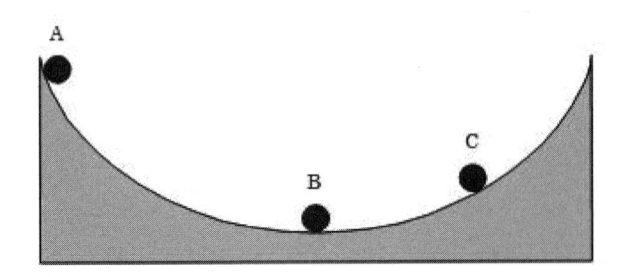

Figure 4: The Isochonism of Cycloids

When the swing of Galileo's pendulum grows large, the isochronism breaks down and the cycle time grows longer. If the pendulum moves back and forth on a cycloid, isochronism should be satisfied. Huygens (1629-1695) implemented an isochronic pendulum as follows. Two cycloids are described ($0 \leq \theta \leq 4\pi$), as shown in Figure 5. When a rope of length $4a$ (the length of the cycloid curve is $8a$) with point B at its center is pulled from point D, the trajectory P forms a cycloid. This kind of curve is known as an involution.

[Question 2] Prove that the cycloid's involution is indeed a cycloid.

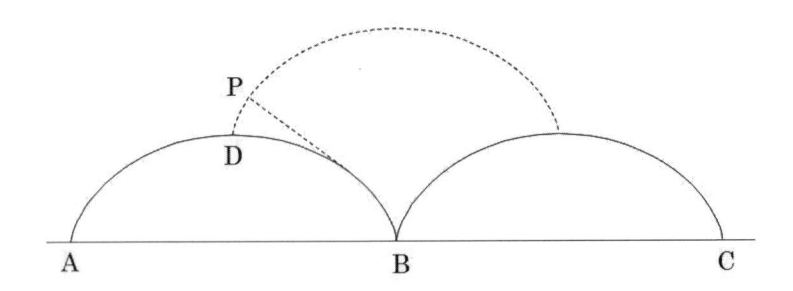

Figure 5: The Cycloid's Involution Is a Cycloid

The cycloid pendulum devised by Huygens is the same as Figure 5 flipped vertically with the central half removed (Figure 6). With this pendulum, even when the swing is large, isochronism is maintained.

Clocks employing this principle are more accurate than Galileo clocks. The length of the pendulum, l, is half the length of the pendulum's swing cycle, $4a$. The period, T is

$$T = 2\pi\sqrt{\frac{4a}{g}}. \tag{7.4}$$

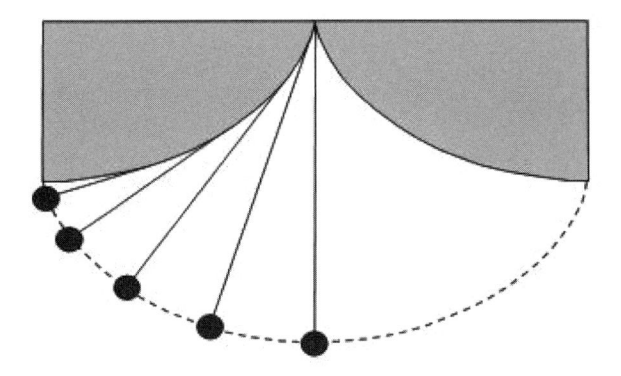

Figure 6: Huygens's Pendulum

Nowadays actual models of the Brachistochrone curve can be seen only in science museums. But we should not forget that the problem of quickest descent mathematically developed the study of the cycloid and the calculus of variations, and contributed to the improvement of pendulums.

Reference

[1] S. Takakuwa, *Bibun Hoteishiki to Henbunho* [*Differential Equations and the Calculus of Variations*], Tokyo: Kyoritsu, (2003).

CHAPTER 24
Machin's Formula and Pi

Abstract: This article explains the calculation of Pi historically, focusing on Machin's formula. Archimedes' formula is shown first, followed by Machin's formula using Gregory's formula. Machin's formula makes particularly ingenious use of the double and quadruple angle trigonometric addition formulae. The chapter closes with an explanation of Takano's formula.

AMS Subject Classification: 68N02, 00A09, 97A20
Key Words: Pi, Machin's formula, Archimedes' method, Gregory's formula, Pythagorean triangles, Takano's formula

1 Finding Pi to 1000 Decimal Places

I've been working on and researching computer-related topics since 1970 - for almost 40 years. I never really had much interest in calculating Pi, the so-called circle ratio. The competition between Japan and America to compute Pi to more and more digits was ongoing, but it was an issue of relevance only to people with access to supercomputers, and I thought it held no interest for the average person. Recently, Prof. Yoshio Kimura sent me an enjoyable book called 'Playing with Simple Computer Programming' [1]. It contains an explanation of the calculation of Pi using Decimal BASIC, according to which, with 50 or so lines of code, Pi could be obtained to 100 decimal places. Recent personal computers have better performance, and computing Pi to 1000 decimal places is now an easy matter.

The equation used here is known as Machin's formula.

$$\frac{\pi}{4} = 4\arctan\frac{1}{5} - \arctan\frac{1}{239}$$

This equation was discovered in 1706 by John Machin, and has been in use for around 300 years. It is suitable for the calculation of Pi because it converges quickly. In this chapter, let's think about how this equation was introduced, and how it can be programmed into a computer.

2   Archimedes' Method

Pi is the ratio of the length of the diameter of a circle to its circumference. At elementary school, we learn its value as 3.14. In practical terms 3 digits are sufficient, but the development of computers and advances in mathematics have worked together to update the value of Pi to a remarkable number of places. First of all, let's take a look at the methods that were adopted before the advent of the calculator. Archimedes, from ancient Greece, calculated Pi using regular polygons which contact and either enclose, or are enclosed by, a given circle. Suppose the radius of a circle is 1 ($OA = OB = 1$). For regular hexagons, the total lengths of the sides of the enclosed hexagon and the enclosing hexagon are 6 and $4\sqrt{3}$ respectively. Thus, $6 < 2\pi < 4\sqrt{3}$ so $3 < \pi < 2\sqrt{3} \approx 3.464$ (Figure 1).

For regular dodecagons (with 12 sides), $OA = OB = OC = 1$ and we may define $AB = a$ and $AC = b$. Denoting the midpoint of AB by M, setting $MC = x$, and applying Pythagoras' theorem to the triangles OAM and ACM, we have

$$1^2 = \left(\frac{a}{2}\right)^2 + (1-x)^2, \quad b^2 = \left(\frac{a}{2}\right)^2 + x^2.$$

Solving this confirms that the circle ratio $\pi > 6b \approx 3.105$ (Figure 2).

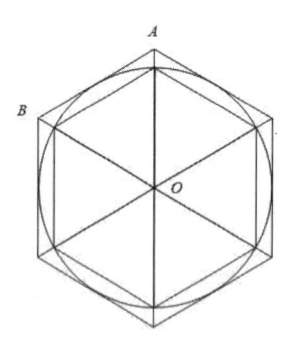

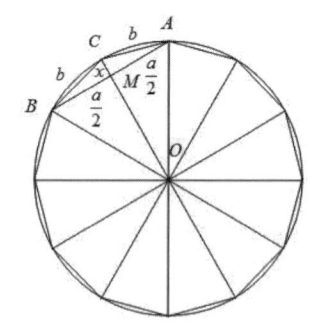

Figure 1: Regular Hexagon

Figure 2: Regular Dodecagon

It is also possible to compute the total side lengths for regular 24-sided and regular 96-sided polygons using Pythagoras' theorem. I'd like for those readers who are interested to confirm this for themselves. In the time of the ancient Greeks, who were not aware of irrational numbers, Archimedes used a 96-sided regular polygon, and calculated the value 3.14.

3 Gregory's Formula

The value was evaluated using regular polygons up until the 17th century. The number of digits was not increased using brute force calculation methods, as these merely increased the number of edges. This reform had to wait until the development of differential Calculus. In 1671, Gregory discovered what is known as Gregory's sequence.

$$\arctan(x) = \sum_{n=0}^{\infty} \frac{(-1)^n}{2n+1} x^{2n+1} = x - \frac{1}{3}x^3 + \frac{1}{5}x^5 - \frac{1}{7}x^7 + \frac{1}{9}x^9 - \cdots$$

In 1674, Leibniz also made the same discovery independently, and the sequence is also called the Gregory-Leibniz sequence. Substituting $x = 1$ in this formula yields the following sequence.

$$\frac{\pi}{4} = 1 - \frac{1}{3} + \frac{1}{5} - \frac{1}{7} + \frac{1}{9} - \frac{1}{11} \cdots$$

The convergence of this sequence is extremely slow, but by taking the partial sum, an approximation to the value of π may be obtained. Let's take a look at how Gregory's sequence was introduced. Since Gregory's sequence is a Taylor expansion of $\arctan(x)$, let's take a look at this as well. For an infinitely differentiable function $f(x)$, the so-called Taylor series is the power series, $\sum_{n=0}^{\infty} \frac{f^{(n)}(a)}{n!}(x-a)^n$, and when this series has the same value as the original function, $f(x)$ is said to have a Taylor expansion. This may be thought of in terms of a neighborhood around $x = a$, and is known as the Taylor expansion around $x = a$. When $a = 0$, the expansion is $\sum_{n=0}^{\infty} \frac{f^{(n)}(a)}{n!} x^n$, this particular case is also called the Maclaurin expansion.

$$f(x) = f(0) + f^{(1)}(0)x + \frac{f^{(2)}(0)}{2!}x^2 + \frac{f^{(3)}(0)}{3!}x^3 + \cdots$$

Leaving the rigorous proof of this equation for another time, isn't it possible to see intuitively that this formula is correct, when both sides are differentiated? tan(tangent) and arctan(arctangent) are mutually inverse functions and may be written as follows.

$$y = \tan x \quad (-\frac{\pi}{2} < x < \frac{\pi}{2})$$

$$x = \tan^{-1} y$$

The notation $y = \arctan x$ is also used. Putting $y = f(x) = \arctan x$, it is clear that,

$$f(0) = 0.$$

Let's think about the derivative of arctan in order to calculate $f^{(1)}(0)$. Differentiating it in this form is difficult, so let's first consider

$$x = \tan y.$$

The derivative of x on the left-hand side is 1, and the right-hand side is a composite function so,

$$1 = (1 + \tan^2 y)\frac{dy}{dx}.$$

Solving this yields the 1st derivative.

$$\frac{dy}{dx} = \frac{1}{1 + \tan^2 y} = \frac{1}{1 + x^2}$$

This can be written

$$f^{(1)}(x) = \frac{1}{1 + x^2}.$$

The 2nd and 3rd derivatives calculated from this equation are as follows.

$$f^{(2)}(x) = \frac{-2x}{(1 + x^2)^2}$$

$$f^{(3)}(x) = \frac{2(3x^2 - 1)}{(1 + x^2)^3}$$

Substituting $x = 0$ into these equations,

$$f^{(1)}(0) = 1, \quad f^{(2)}(0) = 0, \quad f^{(3)}(0) = -2.$$

Substituting these into the terms of the Taylor expansion yields the following equation.

$$\arctan x = x - \frac{x^3}{3} + \frac{x^5}{5} - \frac{x^7}{7} + \cdots$$

4 Machin's Formula

The formula for the circle ratio using the arctangent discovered by the Englishman John Machin in 1706 was mentioned above. Let's restate Machin's formula.

$$\frac{\pi}{4} = 4\arctan\frac{1}{5} - \arctan\frac{1}{239}$$

Machin used Gregory's series with this formula and obtained Pi to 100 decimal places. Allow me to explain the relationship between Machin's formula and Gregory's series.

By substituting $x = \frac{1}{5}$ or alternatively $x = \frac{1}{239}$ into Gregory's series,

$$\arctan\frac{1}{5} = \frac{1}{5} - \frac{1}{3 \cdot 5^3} + \frac{1}{5 \cdot 5^5} - \frac{1}{7 \cdot 5^7} + \cdots$$

$$\arctan\frac{1}{239} = \frac{1}{239} - \frac{1}{3 \cdot 239^3} + \frac{1}{5 \cdot 239^5} - \cdots$$

can be obtained. Substituting these into Machin's formula yields the following.

$$\frac{\pi}{4} = \arctan\frac{1}{5} - \arctan\frac{1}{239}$$
$$= 4 \times (\frac{1}{5} - \frac{1}{3 \cdot 5^3} + \frac{1}{5 \cdot 5^5} - \frac{1}{7 \cdot 5^7} + \cdots) - (\frac{1}{239} - \frac{1}{3 \cdot 239^3} + \frac{1}{5 \cdot 239^5}$$
$$- \frac{1}{7 \cdot 239^7} + \cdots)$$

This is how the circle ratio can be computed.

Now let's look at the details in Machin's formula. This formation of the equation can be explained with the elegant use of trigonometric addition formulae. The addition formula for tan is an old friend that appears in high-school mathematics textbooks.

$$\tan(\alpha + \beta) = \frac{\tan\alpha + \tan\beta}{1 - \tan\alpha\tan\beta}$$

Setting $\alpha = \beta$ in this formula can be used to derive the double angle formula.

$$\tan 2\alpha = \frac{2\tan\alpha}{1 - \tan^2\alpha}$$

Setting $\tan\alpha = \dfrac{1}{5}$ means that $\alpha = \arctan\dfrac{1}{5}$, and the double angle and quadruple angle tangents can be obtained from the formulae above as follows.

$$\tan 2\alpha = \dfrac{2 \times \dfrac{1}{5}}{1 - \dfrac{1}{5} \times \dfrac{1}{5}} = \dfrac{10}{24} = \dfrac{5}{12}$$

$$\tan 4\alpha = \dfrac{2 \times \dfrac{5}{12}}{1 - \dfrac{5}{12} \times \dfrac{5}{12}} = \dfrac{2 \times 5 \times 12}{12 \times 12 - 5 \times 5} = \dfrac{120}{119}$$

$\tan 4\alpha$ is $\dfrac{120}{119}$, which is very close to 1, since it only differs by $\dfrac{1}{119}$. Calculating the tan of the difference between 4α and $\dfrac{\pi}{4}$ reveals the following attractive form.

$$\tan\left(4\alpha - \dfrac{\pi}{4}\right) = \dfrac{\tan 4\alpha - \tan\dfrac{\pi}{4}}{1 + \tan 4\alpha \tan\dfrac{\pi}{4}} = \dfrac{\dfrac{120}{119} - 1}{1 + \dfrac{120}{119} \times 1} = \dfrac{120 - 119}{119 + 120} = \dfrac{1}{239}$$

From this formula,

$$4\alpha - \dfrac{\pi}{4} = \arctan\dfrac{1}{239}$$

and

$$\dfrac{\pi}{4} = 4\arctan\dfrac{1}{5} - \arctan\dfrac{1}{239},$$

thus constituting a derivation of Machin's formula.

Let's think about Machin's formula from another angle. Beginning from $\tan\alpha = \dfrac{1}{5}$ we thought about the fact that $\tan 2\alpha = \dfrac{5}{12}$ and $\tan 4\alpha = \dfrac{120}{119}$. There is a triangle which corresponds to these, and it is shown in Figure 3.

In the double angle case, the edge lengths are in the ratio $12 : 5 : 13$, and in the quadruple angle case they are $119 : 120 : 169$. These values satisfy the relationship expressed by Pythagoras' theorem.

$$13^2 = 12^2 + 5^2, \quad 169^2 = 119^2 + 120^2$$

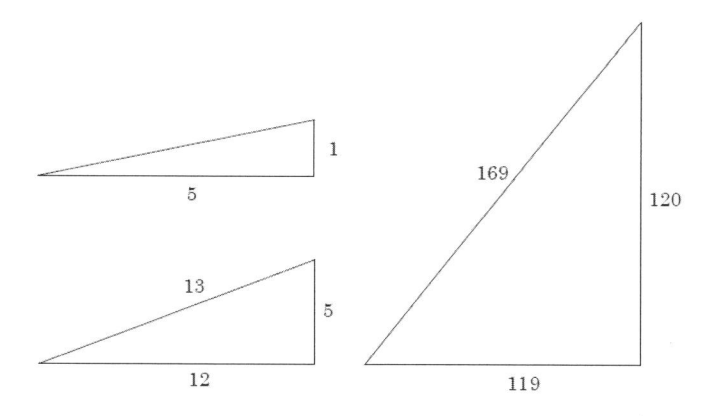

Figure 3: Double and Quadruple Angle Triangles

The double and quadruple angle triangles are also called Pythagorean triangles. Pythagorean triangles satisfy the following relationship.

Theorem
For mutually prime natural numbers m, n, if $\tan \alpha = \dfrac{n}{m}$ $(m > n)$, then the right-angled triangle which has an acute angle 2α is a Pythagorean triangle. Also, the ratio of the lengths of the three edges is $m^2 - n^2 : 2mn : m^2 + n^2$.

Substituting $m = 5$ and $n = 1$ into this theorem, the ratio of the lengths of the edges in the double angle triangle can be calculated from $m^2 - n^2 = 24$, $2mn = 10$ and $m^2 + n^2 = 26$, revealing the ratio $24 : 10 : 26 = 12 : 5 : 13$.

Substituting $m = 12$ and $n = 5$ yields $m^2 - n^2 = 119$, $2mn = 120$ and $m^2 + n^2 = 169$ from which the quadruple angle triangle's edge length ratio can be calculated. Machin's formula makes ingenious use of the tan double and quadruple angle trigonometric addition formulae. There may be many students who wonder whether the trigonometric and multiple angle formulae have much of a role in the mathematics that appears in examinations, but I'd like people to make a point of remembering the tremendous use made of these mathematical assets.

5 Around Machin's Formula

Machin's formula was explained above, but regarding its derivation, just how the formula was discovered seems to be unknown. Perhaps Machin's formula was discovered by accident. Or perhaps it was obtained by building on a mathematical concept. Let's investigate whether there are any other formulae like Machin's. Consider the following series, where a_k represent integers which may be positive or negative, and b_k represent positive integers.

$$\frac{\pi}{4} = \sum_{k=1}^{n} a_k \arctan \frac{1}{b_k}$$

Regarding the values shown in the 2nd term, the following formulae are known.

(1) $\dfrac{\pi}{4} = \arctan \dfrac{1}{2} + \arctan \dfrac{1}{3}$

(2) $\dfrac{\pi}{4} = 2\arctan \dfrac{1}{3} + \arctan \dfrac{1}{7}$

(3) $\dfrac{\pi}{4} = 2\arctan \dfrac{1}{2} - \arctan \dfrac{1}{7}$

(4) $\dfrac{\pi}{4} = 4\arctan \dfrac{1}{5} - \arctan \dfrac{1}{239}$

(1) was conceived by Euler, and can be proven using the tangent addition formula. Try this yourself. The type of problem shown in Figure 4 can sometimes be found in examination reference books. This problem requires one to prove that

$$\alpha = \beta + \gamma,$$

and the original version of this problem seems to be related to Euler's formula for the calculation of Pi.

(4) is Machin's formula. When the values of a_k and b_k are small, it's possible to confirm using calculation by hand, but when the values are large, as in the case of Machin's formula, one must admit defeat. These days we have computers, so comprehensive search methods incorporating a computer program are possible, but it remains a question how Machin's formula was discovered in the 18th century when there were no computers.

When the number of terms is increased to 3, the following forms may be expressed. (6) is Gauss' formula.

(5) $\dfrac{\pi}{4} = \arctan \dfrac{1}{2} + \arctan \dfrac{1}{5} + \arctan \dfrac{1}{8}$

(6) $\dfrac{\pi}{4} = 12\arctan \dfrac{1}{18} + 8\arctan \dfrac{1}{57} - 5\arctan \dfrac{1}{239}$

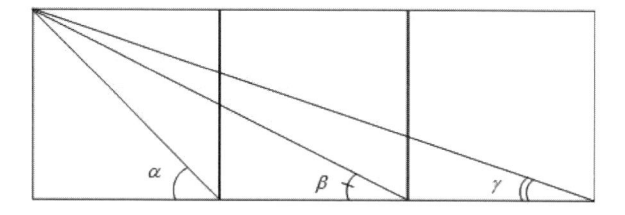

![Figure 4 diagram]

Figure 4: $\alpha = \beta + \gamma$

6 Kikuo Takano's Formula

Now, returning to Machin's formula, let's think about the actual calculation. Even despite knowing Machin's formula, there are some people who can only calculate Pi to 7 significant digits using a computer. Also, given a calculator which can only handle 8 digits, some people think it's only possible to calculate 8 digits. Using a calculator with 8 digits, in principle, it's possible to obtain Pi to 1000 digits, so let's look at the method.

Allow me to explain using $1 \div 239$. For the sake of brevity, let's calculate the digits 3 at a time.

$1 \div 239 = 0$ remainder 1

$1000 \div 239 = 4$ remainder 44

$44000 \div 239 = 184$ remainder 24

$24000 \div 239 = 100$ remainder 100

By arranging these quotient parts we obtain the answer:

0. 004 184 100...

Regarding the recent competition between Japan and America to calculate Pi, in 2002 Yasumasa Kaneda calculated 1 trillion, 2411 hundred million digits using Takano's formula (announced in 1982) on a Hitachi SR8000. See Takano for more information on his formula [2]. With respect to Machin's formula, this formula is incomparably complicated. New formulae are discovered one after the other, but the use of computers is common. However, computers alone are useless, and their foundations are based on many mathematical formulae.

$$\frac{\pi}{4} = 12 \arctan \frac{1}{49} + 32 \arctan \frac{1}{57} - 5 \arctan \frac{1}{239} + 12 \arctan \frac{1}{110443}$$

References

[1] Y. Kimura, *PC wo Asobu: Kantan Programming* [*Enjoying PC: Simple Programming*], Tokyo: Kodansha, (2003).

[2] K. Takano, Pi no Arctangent Relation wo Motomete [Finding the Arctangent Relation of π], *Bit*, 15(4), 83-91, (1983).

CHAPTER 25
Burnside's Lemma

Abstract: There is a famous problem which involves discriminating the faces of a die using 3 colors: how many different patterns can be produced? This article introduces Burnside's lemma which is a powerful method for handling such problems. It requires a knowledge of group theory, but is not so difficult and is likely to be understood by high-school students.

AMS Subject Classification: 19A22, 00A09, 97A20
Key Words: Burnside's counting theorem, Pólya's formula, Permutation, Orbit, Invariant, Equivalence, Equivalence class

1 Discriminating Dice Using 3 Colors

When I set the following problem in a certain magazine, one reader put forward a wonderful solution using Burnside's lemma [3]. I didn't know of Burnside's lemma. It requires a knowledge of group theory and familiarity with the appropriate symbology, but it's not so difficult and high-school students can probably understand it.

The problem I set was as follows. There are a number of squares divided with diagonal borders and colored differently, as shown in Figure 1. How many different possible patterns are there when 4 such squares forming a 2×2 grid are filled in? Cases of symmetrical colors, rotational symmetries and mirror symmetries are regarded as equivalent.

2 Case-by-Case Solution

The objective in this chapter is to introduce Burnside's lemma, so let's begin the explanation with a well-used example. Suppose there is a die like that shown in Figure 2. When the six faces of this die are divided by painting them each with one of three colors, how many different patterns can be produced?

262 CHAPTER 25

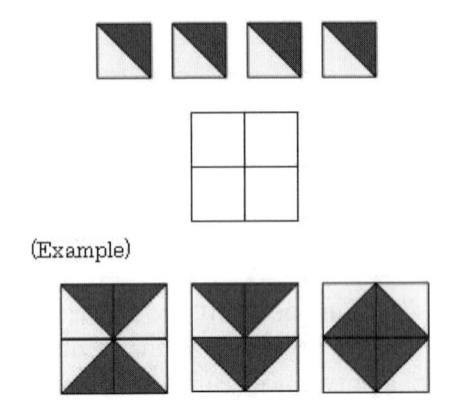

(Example)

Figure 1: How Many Ways in Total Are There to Make Different Patterns?

To begin with, allow me to explain a general case-based solution. Dice are cubes so they have six faces. Suppose they are each painted with one of three colors (say blue, yellow or red). There are three different possible cases for the color of each face. Thus, since each of the faces are independent, in total there are
$$3^6 = 729$$
ways of coloring the faces.

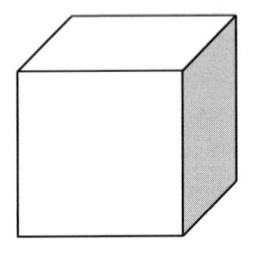

Figure 2: Discriminating the Faces of a Die Using 3 Colors

Checking all of these is a laborious task. Let's therefore try thinking about the cases organized in the following way.

Classifying according to how many colors are used yields three cases: 1 color, 2 colors and 3 colors. Let's attempt a top-level classification on this basis. When only 1 color is used, there are 3 cases, *i.e.*, when all 6 faces are simply either blue, yellow or red.

Next, in the case that 2 colors are used, the ratio of the colors can take 3 different values, 5 : 1 faces, 4 : 2 or 3 : 3. Let's use this as a mid-level classification. There is a further relationship according to which 2 of the 3 colors are chosen. When all 3 colors are used, there are 3 possible ratios of the colors, 4 : 1 : 1 faces, 3 : 2 : 1 or 2 : 2 : 2. Counting up the patterns in this way, there are 57 different cases.

This counting operation is probably impossible with pencil and paper. I actually drew a net of the cube on a computer, and checked the arrangements of the colors over and over again. It turned out that many times the patterns that I had imagined to be different inside my head were actually the same. In the end I bought a wooden block (with 2 cm edges) from the carpentry section of a DIY store, and sticking colored paper on the faces, confirmed the 57 patterns. Figure 4 shows each of the 57 patterns. Blue is represented by B, yellow by Y and red by R, and the numbers 1 to 6 in the header row correspond to the numbers of the faces in the net (Figure 3).

3 Solution Using Burnside's Lemma

No matter how cautiously the equations enumerating the cases are counted up, counting errors and oversights are sometimes bound to happen. For situations like this, there is a powerful method which applies knowledge from group theory known as Burnside's lemma. I'll explain below.

Burnside's lemma is described as follows in the free encyclopedia, Wikipedia [1]. Burnside's lemma is also known as Burnside's counting theorem, Pólya's formula, the Cauchy-Frobenius lemma and the orbit-counting theorem. These all refer to the same thing. Burnside wrote down this lemma in 1900. According to the history of mathematics, Cauchy wrote it in 1845 and Frobenius in 1887, so Burnside was not the first person to discover it, and some people refer to it formally as 'not-Burnside's lemma.'

When a permutation group G is applied to a set X, if the number of elements which are invariant under an element g of the group G is denoted X^g, then the number of orbits, $|X/G|$, is given by the following formula.

$$|X/G| = \frac{1}{|G|} \sum_{g \in G} |X^g|$$

The number of orbits means the number of things which are equivalent.

It was shown above that there are $3^6 = 729$ different ways of partitioning the 6 faces of a die using 3 colors. This set is denoted X. There

264 CHAPTER 25

```
   1
5  2  6
   3
   4
```

Figure 3: The Numbers Corresponding to the Faces on the Die

No.	1	2	3	4	5	6	B	Y	R		
1	B	B	B	B	B	B	6				
2	Y	Y	Y	Y	Y	Y		6		6	1 color
3	R	R	R	R	R	R			6		
4	B	Y	B	B	B	B	5	1			
5	B	R	B	B	B	B	5		1		
6	Y	B	Y	Y	Y	Y	1	5			
7	Y	R	Y	Y	Y	Y		5	1	5+1	
8	R	B	R	R	R	R	1		5		
9	R	Y	R	R	R	R		1	5		
10	B	Y	Y	B	B	B	4	2			
11	B	R	R	B	B	B	4		2		
12	Y	B	B	Y	Y	Y	2	4			
13	Y	R	R	Y	Y	Y		4	2		
14	R	B	B	R	R	R	2		4		
15	R	Y	Y	R	R	R		2	4	4+2	2 colors
16	B	Y	B	Y	B	B	4	2			
17	B	R	B	R	B	B	4		2		
18	Y	B	Y	B	Y	Y	2	4			
19	Y	R	Y	R	Y	Y		4	2		
20	R	B	R	B	R	R	2		4		
21	R	Y	R	Y	R	R		2	4		
22	B	Y	Y	B	B	Y	3	3			
23	Y	R	R	Y	Y	R		3	3		
24	R	B	B	R	R	B	3		3	3+3	
25	B	Y	Y	Y	B	B	3	3			
26	Y	R	R	R	Y	Y		3	3		
27	R	B	B	B	R	R	3		3		
28	B	Y	R	B	B	B	4	1	1		
29	Y	R	B	Y	Y	Y	1	4	1		
30	R	B	Y	R	R	R	1	1	4	4+1+1	
31	B	Y	B	R	B	B	4	1	1		
32	Y	R	Y	B	Y	Y	1	4	1		
33	R	B	R	Y	R	R	1	1	4		
34	B	Y	Y	R	B	B	3	2	1		
35	Y	R	R	B	Y	Y	1	3	2		
36	R	B	B	Y	R	R	2	1	3		
37	B	R	R	Y	B	B	3	1	2		
38	Y	B	B	R	Y	Y	2	3	1		
39	R	Y	Y	B	R	R	1	2	3		
40	B	Y	R	B	B	Y	3	2	1		
41	Y	R	B	Y	Y	R	1	3	2		
42	R	B	Y	R	R	B	2	1	3	3+2+1	3 colors
43	B	R	Y	B	B	R	3	1	2		
44	Y	B	R	Y	Y	B	2	3	1		
45	R	Y	B	R	R	Y	1	2	3		
46	B	Y	R	Y	B	B	3	2	1		
47	Y	R	B	R	Y	Y	1	3	2		
48	R	B	Y	B	R	R	2	1	3		
49	B	R	Y	R	B	B	3	1	2		
50	Y	B	R	B	Y	Y	2	3	1		
51	R	Y	B	Y	R	R	1	2	3		
52	Y	B	R	B	R	Y	2	2	2		
53	R	Y	B	Y	B	R	2	2	2		
54	B	R	Y	R	Y	B	2	2	2	2+2+2	
55	B	Y	R	B	Y	Y	2	2	2		
56	B	Y	R	R	Y	B	2	2	2		
57	B	R	B	R	Y	Y	2	2	2		

Figure 4: Discriminating the Die Faces Using 3 Colors (B: Blue, Y: Yellow, R: Red)

are 4 types of rotation group G which can be considered with respect to X.

(1) Rotation by 90° about an axis through two parallel faces (this can be performed in 6 different ways). In the case of a rotation of 90° about the axis through faces ABFE and DCGH shown in Figure 5(a), the parallel faces ABFE and DCGH may be different colors, so there are 3^2 different colorings of these faces, but the four faces which are moved by 90°, ABCD, BFGC, EFGH and AEHD must all be the same color, so there are 3 colorings. For each axis there are thus 3^3 possibilities, which gives a total of 6×3^3 possibilities.

(2) Rotation by 180° about an axis through two parallel faces (this can be performed in 3 different ways). In the case of a rotation of 180° about the axis through faces ABFE and DCGH shown in Figure 5(a), the parallel faces ABFE and DCGH may be different colors, so there are 3^2 colorings of these faces. Since the rotation is by 180°, it is necessary for the corresponding faces, among the four which remain, to have the same color. For example, faces ABCD and EFGH must be the same, as well as BFGC and AEHD. This gives 3^2 possible colorings. There are thus 3^4 possibilities for each axis, which gives a total of 3×3^4 possibilities.

(3) Rotation by 120° about an axis through two opposite vertices (this can be performed in 8 possible ways). In the case of a rotation of 120° about the axis through vertices B and H shown in Figure 5(b), the 3 faces adjacent to vertex B (ABCD, BFGC and ABFE) must all be the same color. Likewise, the 3 faces adjacent to vertex H (DCGH, EFGH and AEHD) must all be the same color. For each axis there are 3^2 combinations of colors, which gives a total of 8×3^2 possibilities.

(4) Rotation by 180° about an axis through two opposite edges (this can be performed in 6 possible ways). In the case of a rotation of 180° about the axis through edges BF and DH shown in Figure 5(c), the 2 faces adjacent to the edge BF (BFGC and ABFE) must be the same color, and the 2 faces adjacent to DH (DCGH and AEHD) must be the same color. Also, the two opposite faces, ABCD and EFGH, which are shifted through 180° must also be the same color. For each axis there are 3^3 combinations of colors, which gives a total of 6×3^3 possibilities.

The number of elements in the rotation group G, including the identity transformation e, is $1+6+3+8+6 = 24$. Applying the information above yields the equation

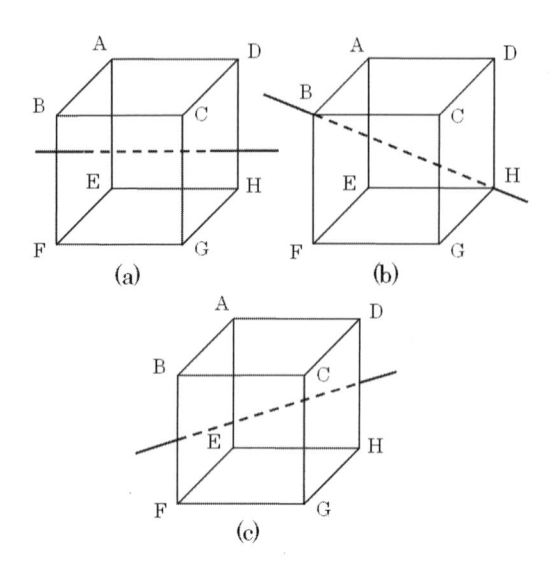

Figure 5: Rotation Axes and Rotation Groups

$$\frac{1}{24}(3^6 + 6 \times 3^3 + 3 \times 3^4 + 8 \times 3^2 + 6 \times 3^3) = 57,$$

and there are thus 57 different patterns.

4 Group Theory, Permutation Groups and Equivalence Classes

Considering a set X, and a permutation group G which acts on the set X, we'd like to obtain the number of equivalence classes in X according to the equivalence relation on X derived from G. This problem can be solved directly by finding the equivalence relation, and then counting the number of equivalence classes. However, when the set X has a particularly large number of elements, such a counting method may be sufficiently awkward as to be beyond human capability.

The number of equivalence classes can be found with Burnside's theory, by counting the numbers of elements (permutations) of X that are invariant under the group. If a given permutation transforms a given element onto itself, then the element is described as 'invariant' under the permutation [2].

The number of elements (permutations) included in the permutation group G is denoted by $|G|$. For a permutation $\pi \in G$, the elements which are mapped by π onto themselves are known as 'invariant,' *i.e.*, they do not vary from their original values, and the number of invariant elements is denoted by $n(\pi)$ [4].

Theorem (Burnside)
For a set X and permutation group G, the number of equivalence classes in X under the equivalence relation imposed by G, written $N(X)$, is given by the following formula.

$$N(X) = \frac{1}{|G|} \sum_{\pi \in G} n(\pi) \qquad (1)$$

A simple example is shown below. Denote the 3 vertices of an equilateral triangle, such as that shown in Figure 6, by A, B and C, and consider the cases when these vertices are colored either red or white. The total number of ways of coloring the vertices, as shown by $P = \{P_1, P_2, \cdots, P_8\}$ in Figure 7, is $2^3 = 8$. This triangle may, by way of example, be subjected to rotations of 120° in a clockwise direction about an axis perpendicular to the triangle and passing through its centre. This transforms P_2 in Figure 7 to P_3, and P_3 to P_4. Sets like P_2, P_3, P_4 can thus be considered "equivalent."

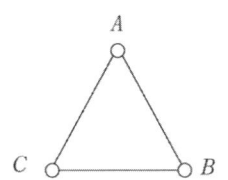

Figure 6: Equilateral Triangle

The permutations of $P_1, \cdots, P_8$ that result when the equilateral triangle is rotated by 120° or 240° in a clockwise direction about a perpendicular axis passing through its center can be expressed as shown below.

$$\pi_1 = (P_1)(P_2 P_3 P_4)(P_5 P_6 P_7)(P_8)$$
$$\pi_2 = \pi_1 \pi_1 = (P_1)(P_2 P_4 P_3)(P_5 P_7 P_6)(P_8)$$

The number of invariant elements for each permutation π, written $n(\pi)$, is given by the following equation.

$$n(\pi_1) = n(\pi_2) = 2 \qquad (2)$$

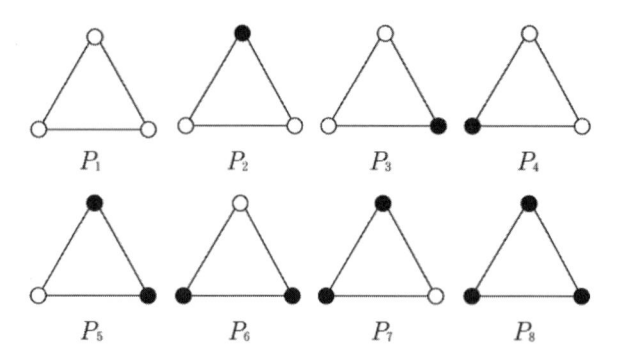

Figure 7: Coloring an Equilateral Triangle

On the other hand, when the triangle is rotated by 180° about an axis from one vertex to the mid-point of the opposite edge, P_5 becomes P_7, or alternatively, P_5 becomes P_6. This reveals that these are indeed "equivalent." For these cases the permutations of $P = \{P_1, P_2, \cdots, P_8\}$ may be expressed as shown below.

$$\pi_3 = (P_1)(P_2)(P_3P_4)(P_5P_7)(P_6)(P_8)$$
$$\pi_4 = (P_1)(P_2P_4)(P_3)(P_5P_6)(P_7)(P_8)$$
$$\pi_5 = (P_1)(P_2P_3)(P_4)(P_5)(P_6P_7)(P_8)$$

The number of invariant elements $n(\pi)$, for each permutation π, is as follows.

$$n(\pi_3) = n(\pi_4) = n(\pi_5) = 4 \tag{3}$$

Considering the permutations $\{\pi_1, \pi_2, \cdots, \pi_5\}$ expressed above, and in addition, the identity permutation which maps every element to itself,

$$\pi_0 = (P_1)(P_2)(P_3)(P_4)(P_5)(P_6)(P_7)(P_8),$$

it can be seen that they constitute a group. In this way, the number of equivalence classes $N(X)$ imposed by the permutation group on the set $P = \{P_1, P_2, \cdots, P_8\}$ is given by the following formula, based on Equation 1, and using Equations (2) and (3) and π_0.

$$N(P) = \frac{1}{|G|} \sum_{i=0}^{5} n(\pi_i) = \frac{1}{6}(8 + 2 \times 2 + 3 \times 4) = 4$$

The number of equivalence classes is thus 4, and it can be seen that the equivalence classes are $\{P_1\},\{P_2,P_3,P_4\},\{P_5,P_6,P_7\}$ and $\{P_8\}$. The transformations of the elements are shown in Table 1, the invariant elements are shown in Table 2 and Table 3 shows the equivalence relations.

	π_0	π_1	π_2	π_3	π_4	π_5
P_1	P_1	P_1	P_1	P_1	P_1	P_1
P_2	P_2	P_3	P_4	P_2	P_4	P_3
P_3	P_3	P_4	P_2	P_4	P_3	P_2
P_4	P_4	P_2	P_3	P_3	P_2	P_4
P_5	P_5	P_6	P_7	P_7	P_6	P_5
P_6	P_6	P_7	P_5	P_6	P_5	P_7
P_7	P_7	P_5	P_6	P_5	P_7	P_6
P_8	P_8	P_8	P_8	P_8	P_8	P_8

Table 1. Transformations of the Elements According to the Permutations

	π_0	π_1	π_2	π_3	π_4	π_5
P_1	=	=	=	=	=	=
P_2	=		=			
P_3	=				=	
P_4	=					=
P_5	=					=
P_6	=			=		
P_7	=				=	
P_8	=	=	=	=	=	=
	8	2	2	4	4	4

Table 2. Invariant Elements

	P_1	P_2	P_3	P_4	P_5	P_6	P_7	P_8
P_1	√							
P_2		√	√	√				
P_3		√	√	√				
P_4		√	√	√				
P_5					√	√	√	
P_6					√	√	√	
P_7					√	√	√	
P_8								√

Table 3. Equivalence Relations

References

[1] Burnside's lemma, from Wikipedia.
http://en.wikipedia.org/wiki/Burnside's_lemma.

[2] C.L. Liu, (translated by H. Narishima, and J. Akiyama) *Elements of Discrete Mathematics*, 2nd Edition, Tokyo: Ohmsha, 450-457, (1995).

[3] Y. Nishiyama, Elegant na Kaito wo Motomu [Seeking Elegant Solutions], *Sugaku Semina* [*Mathematics Seminar*], 45(9), 95-100, (2006).

[4] T. Oyama, *Power-up Risan Sugaku* [*Power-up Discrete Mathematics*], Tokyo: Kyoritsu, 70-76, (1997).

CHAPTER 26
Gauss' Method of Constructing a Regular Heptadecagon

Abstract: This article explains how to construct a regular heptadecagon according to the theory of cyclotonic equations which was discovered by Gauss in 1796. The author also shows how to construct any root or fraction.

AMS Subject Classification: 20A02, 00A09, 97A20
Key Words: Constructing a regular heptadecagon, Theory of cyclotonic equations, Modulo, Prime number and primitive root, Constructing roots and fractions

1 Introduction

We know about historically famous theories in mathematics, but there are many cases where we don't know the proofs. Fermat's last theorem, Galois' theorem, Gödel's incompleteness theorem..., once you start, there's nowhere to stop. Until recently I did not know the proof supporting Gauss' method for constructing a regular heptadecagon - a polygon with 17 sides. The construction method for an arbitrary regular n-sided polygon may be explained as follows, according to the *Dictionary of Mathematics* (Iwanami Shoten Publishing). The necessary and sufficient conditions for a regular n-sided polygon to be possible to construct were established by Gauss. Given the prime factorization of n,

$$n = 2^\lambda P_1 \cdots P_k \quad (\lambda \geq 0) \tag{1}$$

all of $P_1, \cdots, P_k$ must be distinct primes taking the form $2^h + 1$ (Fermat numbers). Inserting values for λ and h into this formula, the following values may be obtained for n.

$$n = 3, 4, 5, 6, 8, 10, 12, 15, 17, \cdots$$

When I reproduced the following quote from a certain document, "It is well-known that Gauss obtained a geometrical method for constructing a regular heptadecagon," I received the query, "By the way, how do you construct the regular heptadecagon?" I had copied verbatim from the dictionary and didn't know the answer. The construction methods for regular triangles and regular pentagons were established by Euclid, the ancient Greek. However, considering whether our generation has completely mastered even the construction of regular pentagons, it is doubtful. For the regular pentagon, there are two construction methods. One begins by establishing a single edge, while the other contacts a circle internally. Neither requires a protractor, and both may be drawn using only a pair of compasses and a ruler. Since it's not the main topic of this chapter the explanation is abbreviated, but it utilizes the facts that

$$\cos\frac{\pi}{5} = \frac{1+\sqrt{5}}{4} \tag{2}$$

and

$$\sin\frac{\pi}{5} = \frac{\sqrt{10-2\sqrt{5}}}{4}. \tag{3}$$

I am not an algebra specialist, so in the explanation that follows I'd like for the reader to understand the construction method for the regular heptadecagon attempted by Gauss as far as I have investigated it.

2 From Gauss' Diary

Chapter 1 of Takagi Teiji's *History of Modern Mathematics* reports that when on the 30th March 1796, the 19-year-old Gauss opened his eyes and arose from his bed, a method for the outstanding problem of constructing a regular heptadecagon occurred to him and was thus recorded in his diary. An outline of the method goes as follows [3].

If it is only required to prove the possibility of constructing a regular heptadecagon, the solution is clear and simple. Taking

$$360° = 17\phi,$$

if the value of $\cos\phi$ may be expressed as a square root, then it is possible to construct the figure. $\cos\phi$ represents the x coordinate of a point on the circumference of a unit circle. Gauss demonstrated the computational process. Let's build on the explanation by looking at the method.

Firstly, Gauss made the following definitions.

$$\cos\phi + \cos 4\phi = a$$
$$\cos 2\phi + \cos 8\phi = b$$
$$\cos 3\phi + \cos 5\phi = c$$
$$\cos 6\phi + \cos 7\phi = d$$

The thing to pay attention to here is the substitution of the parameters a, b, c, d for the values of $\cos\phi$ to $\cos 8\phi$. Perhaps this kind of substitution is only applied to this problem. If these are combined at random, then in total there are $_8C_2 \times {}_6C_2 \times {}_4C_2 = 2520$ possibilities. Do you think Gauss might have investigated every case? He didn't. We will look at this issue in detail later, and it is closely related to the theory of cyclotomic equations.

Next, setting
$$a + b = e$$
$$c + d = f$$

then, as is widely known,

$$[1] \quad e + f = -\frac{1}{2}.$$

In order to understand Equation [1], it's sufficient to remember a problem which often comes up in university entrance examinations.

[Theory] For natural number n, setting

$$S_n = \cos\phi + \cos 2\phi + \cdots + \cos n\phi$$

implies that
$$2 S_n \sin\frac{\phi}{2} = \sin\frac{2n+1}{2}\phi - \sin\frac{\phi}{2}.$$

The proof of this theory involves the application of the product $\to$ sum formula for $2\cos k\phi \sin\frac{\phi}{2}$ ($k = 1, \cdots, n$), so that the intermediate term vanishes and the formula is simplified. Then, substituting $n = 8, \phi = \frac{2\pi}{17}$ into

$$S_n = (\sin\frac{2n+1}{2}\phi - \sin\frac{\phi}{2})/2\sin\frac{\phi}{2}$$

yields $\sin\dfrac{2n+1}{2}\phi = 0$. Therefore, $S_8 = -\dfrac{1}{2}$.

The products formed by each pair among a, b, c, d may now be obtained. By means of a simple calculation, and taking note of the fact that $\cos n\phi = \cos(17-n)\phi$, the results are

$$\begin{aligned}
2ab &= e + f = -\frac{1}{2}\\
2ac &= 2a + b + d\\
2ad &= b + c + 2d\\
2bc &= a + 2c + d\\
2bd &= a + 2b + c\\
2cd &= e + f = -\frac{1}{2}.
\end{aligned}$$

Looking at one of the equations above, for example, $2ab$, it is obtained as follows.

$2ab = 2(\cos\phi + \cos 4\phi)(\cos 2\phi + \cos 8\phi)$

$= 2\cos\phi\cos 2\phi + 2\cos\phi\cos 8\phi + 2\cos 4\phi\cos 2\phi + 2\cos 4\phi\cos 8\phi$

$= (\cos 3\phi + \cos\phi) + (\underline{\cos 9\phi} + \cos 7\phi) + (\cos 6\phi + \cos 2\phi) + (\underline{\cos 12\phi} + \cos 4\phi)$

Here the substitution $\cos 9\phi = \cos 8\phi, \cos 12\phi = \cos 5\phi$ is made, and the result rearranged, yielding

$= (\cos\phi + \cos 4\phi) + (\cos 2\phi + \cos 8\phi) + (\cos 3\phi + \cos 5\phi) + (\cos 6\phi + \cos 7\phi)$

$= a + b + c + d$

$= e + f$

$= -\dfrac{1}{2}$

ergo,

$$2ac + 2ad + 2bc + 2bd = 4a + 4b + 4c + 4d,$$

i.e.,

$$2ef = -2,$$

or alternatively,

[2] $ef = -1$.

The following is a solution method utilizing the relationship between the solutions and factors of 2nd order equations. In particular, from [1] and [2], the equations for e and f are the roots of

$$x^2 + \frac{1}{2}x - 1 = 0.$$

One is thus
$-\frac{1}{4} + \sqrt{\frac{17}{16}}$, while the other is $-\frac{1}{4} - \sqrt{\frac{17}{16}}$.
A glance is sufficient to reveal from their values that the first is e, and the second is f.

Now, the following equation has roots a and b.

$$x^2 - ex - \frac{1}{4} = 0.$$

The values of the roots are

$$\frac{1}{2}e \pm \sqrt{\frac{1}{4} + \frac{1}{4}e^2} = -\frac{1}{8} + \frac{1}{8}\sqrt{17} \pm \frac{1}{8}\sqrt{34 - 2\sqrt{17}}.$$

It is clear here that a is the upper (positive) sign and b is the lower (negative) sign. The reason is that, trivially,

$$a - b = (\cos\phi - \cos 2\phi) + (\cos 4\phi - \cos 8\phi),$$

so in exactly the same way,

$$c = -\frac{1}{8} - \frac{1}{8}\sqrt{17} + \frac{1}{8}\sqrt{34 + 2\sqrt{17}},$$

and

$$d = -\frac{1}{8} - \frac{1}{8}\sqrt{17} - \frac{1}{8}\sqrt{34 + 2\sqrt{17}}.$$

Now finally, $\cos\phi$ and $\cos 4\phi$ are clearly the roots of the following 2nd order equation (because the product $\cos\phi \cdot \cos 4\phi = \frac{1}{2}c$).

$$x^2 - ax + \frac{1}{2}c = 0$$

Consequently,

$$\cos\phi = +\frac{1}{2}a + \sqrt{\frac{1}{4}a^2 - \frac{1}{2}c},$$

and

$$\cos 4\phi = +\frac{1}{2}a - \sqrt{\frac{1}{4}a^2 - \frac{1}{2}c}.$$

Rearranging however yields,

$$2a^2 = 2 + b + 2c$$

so,

$$\cos \phi = \frac{1}{2}a + \sqrt{\frac{1}{4} + \frac{1}{8}b - \frac{1}{4}c} = -\frac{1}{16} + \frac{1}{16}\sqrt{17} + \frac{1}{16}\sqrt{34 - 2\sqrt{17}}$$
$$+ \frac{1}{8}\sqrt{17 + 3\sqrt{17} - \sqrt{34 - 2\sqrt{17}} - 2\sqrt{34 + 2\sqrt{17}}},$$

which is the value Gauss obtained for $\cos \phi$.

3 Gauss' Theory of Cyclotomic Equations

We have seen how the value of $\cos \phi$ needed for the construction of a regular heptadecagon can be obtained, but this calculation was just a confirmation. So far we have not even touched upon the essential question of why it turns out as it does, *i.e.*, the reason for the substitution of the four parameters a, b, c, d for the terms from $\cos \phi$ to $\cos 8\phi$ must be stated.

The reason is not written in Gauss' diary. In order to find out, one must enlist the help of Kurata Reijirou's *Gauss' Theory of Cyclotonic Equations* [2] or *Gauss' Theory of Numbers* translated by Takase Masahito [1]. Think about the following equation.

$$x^n - 1 = 0 \qquad (4)$$

It goes without saying that the roots of this equation are the nth roots of 1, and as is widely known,

$$e^{\frac{2\pi k i}{n}} \quad (k = 0, 1, 2, \cdots, n-1).$$

This has a relationship with Euler's equation.

$$e^{i\theta} = \cos \theta + i \sin \theta \qquad (5)$$

Among the n roots, those which first equal 1 when raised to the power n are known as the primitive nth roots of unity. $e^{\frac{2\pi i}{n}}$ is a primitive nth root of unity. This is a point with angle $\dfrac{2\pi}{n}$ on a unit circle in the complex plane, i.e., it expresses an nth equal part of a complete circumference. $e^{\frac{2\pi ki}{n}}$ expresses the point on the circumference with k-times the angle.

Through the intermediate agency of the complex plane, Equation (4) is tied to the n-sided regular polygon figures originating in ancient Greece. This was Gauss' underlying perspective, and he was the first mathematician to introduce the complex plane.

Now, Equation (4) is

$$x^n - 1 = (x-1)(x^{n-1} + x^{n-2} + \cdots + x + 1),$$

so excluding 1, all the roots of Equation (4) are roots of the following function.

$$F(x) = x^{n-1} + x^{n-2} + \cdots + x + 1 \quad (6)$$

This is known as a cyclotomic equation, or alternatively, as a circle-partitioning equation.

In $x^n - 1 = 0$, if a primitive nth root of unity, denoted ω, is already known, then the solutions of this equation are

$$1, \omega, \omega^2, \cdots, \omega^{n-1}.$$

For example, thinking about $x^3 - 1 = 0$,

$$x^3 - 1 = (x-1)(x^2 + x + 1) = 0,$$

so the 3rd root is 1, $\frac{-1 \pm \sqrt{3}i}{2}$. Taking $\frac{-1+\sqrt{3}i}{2} = \omega$, we have $\frac{-1-\sqrt{3}i}{2} = \omega^2$, and it can be seen that the roots are $1, \omega$ and ω^2. In this case, $3\theta = 2\pi$ so

$$\omega = \cos\theta + i\sin\theta = e^{i\theta}$$
$$\omega^2 = \cos 2\theta + i\sin 2\theta = e^{i2\theta}$$
$$\omega^3 = \cos 3\theta + i\sin 3\theta = e^{i3\theta} = 1$$

and it can be seen that the 3 roots cycle.

At this point I'd like to introduce the following theorem which utilizes the concept of a remainder.

[Theorem] When p is a prime number, the set of roots of the cyclotomic equation

$$F(x) = x^{n-1} + x^{n-2} + \cdots + x + 1$$

may be written as Ω, and apart from 1, all the elements $r \in \Omega$ (which satisfy $x^n - 1 = 0$) are complex numbers. Furthermore, for a positive or negative integer e which is not divisible by p, the following are satisfied.

(1) $r^p = 1, r^{2p} = r^{3p} = \cdots = 1, r^{ep} = 1$

(2) for integers λ and μ

$$\lambda \equiv \mu \pmod{p} \Leftrightarrow r^\lambda = r^\mu$$

(3) for $r \in \Omega$

$$\Omega = \{r^e, r^{2e}, \cdots, r^{e(p-1)}\}, \quad r^e + r^{2e} + \cdots + r^{e(p-1)} = -1$$

Gauss defined a parameter with a value representing an f-step cycle.

[Definition] For an odd prime p, and a primitive pth root of unity, denoted r, and taking $p - 1 = fe$, g as a primitive root of p and λ as an arbitrary integer, the value of the f-step cycle, denoted (f, λ), is defined as follows.

$$(f, \lambda) = [\lambda] + [\lambda h] + [\lambda h^2] + \cdots + [\lambda h^{f-1}] \quad \text{(note that } h = g^e\text{)}$$

At this point let's perform the calculation for a regular heptadecagon. Taking 3 as a primitive root of the prime number 17,

$$p = 17, p - 1 = 16 = 16 \times 1 = f \times e, \quad g = 3, \quad h = g^e = 3^1 = 3,$$

and the 16-step cycle is as follows.

$$(16, 1) = [1] + [3] + [9] + [10] + [13] + [5] + [15] + [11] + [16] + [14]$$
$$+ [8] + [7] + [4] + [12] + [2] + [6]$$

That is to say, when 3 is taken as the primitive root for $\{[1], \cdots, [16]\}$, the result is a cyclic group modulo 17. For example, the 4th term on the right-hand side of the equation above, [10], can be obtained as

$$\lambda h^3 = 1 \times 3^3 = 27 = 10 \pmod{17}.$$

An explanation of primitive roots, modulo and cyclic groups is deferred to books specialized in algebra, number theory, groups and so on.

At any rate, primes and primitive roots have a truly ingenious relationship. I first heard of the term 'primitive root' when I studied subroutines for generating pseudorandom numbers on computers. As a consequence, computers can be made to generate randomly ordered sequences of all the integers they are capable of expressing. For the problem in question, the significant point is that the 16 roots can be reordered according to the concept of remainders. This was how the

16-step cycle could be obtained. Gauss presented a theory decomposing an f-step cycle.

[Theorem] When $p - 1 = abc$, the bc-step cycle (bc, λ) is the sum of b c-step cycles.

$$(bc, \lambda) = (c, \lambda) + (c, \lambda g^a) + (c, \lambda g^{2a}) + \cdots + (c, \lambda g^{a(b-1)})$$

Let's attempt a decomposition of the 16-step cycle into two 8-step cycles as described on p27-28 of Kurata's work [2]. From $p = 17, p - 1 = 16 = 1 \times 2 \times 8 = a \times b \times c$,

$(16, 1) = (8, 1) + (8, 3)$.

Defining $(8, 1)$ and $(8, 3)$ we have the following equations.

$$\begin{aligned}
(8,1) &= [1] + [9] + [9^2] + [9^3] + [9^4] + [9^5] + [9^6] + [9^7] \\
&= [1] + [9] + [81] + [729] + [6561] + [59049] + [531441] + [4782969] \\
&= [1] + [9] + [13] + [15] + [16] + [8] + [4] + [2] \pmod{17} \\
&= [1] + [2] + [4] + [8] + [9] + [13] + [15] + [16] \\
(8,3) &= [3] + [3 \cdot 9] + [3 \cdot 9^2] + [3 \cdot 9^3] + [3 \cdot 9^4] + [3 \cdot 9^5] + [3 \cdot 9^6] + [3 \cdot 9^7] \\
&= [3] + [27] + [243] + [2187] + [19683] + [177147] + [1594323] \\
&\quad + [14348907] \\
&= [3] + [10] + [5] + [11] + [14] + [7] + [12] + [6] \pmod{17} \\
&= [3] + [5] + [6] + [7] + [10] + [11] + [12] + [14]
\end{aligned}$$

Here, the 16-step cycle $(16, 1)$ is decomposed into two 8-step cycles, $(8, 1)$ and $(8, 3)$. If the decomposition theory is applied repeatedly, the decomposition can be continued down to a final 1-step cycle. Figure 1 shows a breakdown of the whole decomposition process. The parameters a, b, c, d, e, f and values of $\cos \phi$ to $\cos 8\phi$ recorded in Gauss' diary are compiled in this figure. Can you understand the reason why Gauss chose the values

$\cos \phi + \cos 4\phi = a, \quad \cos 3\phi + \cos 5\phi = c,$
$\cos 2\phi + \cos 8\phi = b, \quad \cos 6\phi + \cos 7\phi = d$
$a + b = e, \quad c + d = f$

by means of this result?

For $p = 17$, the 16 roots besides 1 may be calculated as shown in Table 1, in a similar way. Figure 2 shows a diagram of a regular heptadecagon. By means of such a decomposition, it becomes possible to factorize the 2nd term on the right-hand side of the equation.

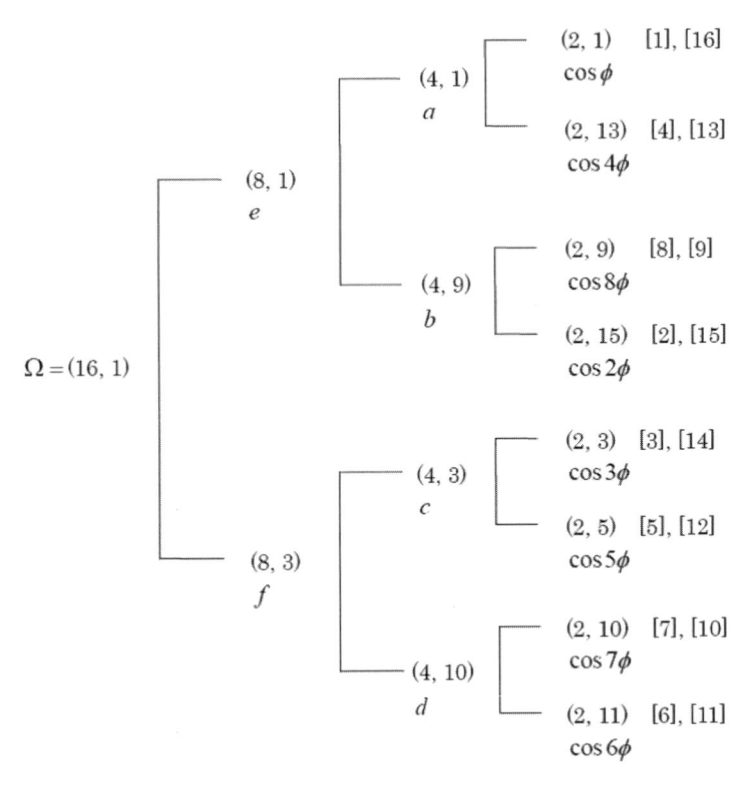

```
                                                     ┌─ (2, 1)   [1], [16]
                                                     │  cos φ
                                        ┌─ (4, 1)  ──┤
                                        │    a       │
                                        │            └─ (2, 13)  [4], [13]
                                        │               cos 4φ
                         ┌─ (8, 1) ─────┤
                         │    e         │            ┌─ (2, 9)   [8], [9]
                         │              │            │  cos 8φ
                         │              └─ (4, 9)  ──┤
                         │                   b       │
                         │                           └─ (2, 15)  [2], [15]
                         │                              cos 2φ
    Ω = (16, 1) ─────────┤
                         │                           ┌─ (2, 3)   [3], [14]
                         │                           │  cos 3φ
                         │              ┌─ (4, 3)  ──┤
                         │              │    c       │
                         │              │            └─ (2, 5)   [5], [12]
                         │              │               cos 5φ
                         └─ (8, 3) ─────┤
                              f         │            ┌─ (2, 10)  [7], [10]
                                        │            │  cos 7φ
                                        └─ (4, 10) ──┤
                                             d       │
                                                     └─ (2, 11)  [6], [11]
                                                        cos 6φ
```

Figure 1: $x^{17} - 1 = (x - 1)(x^{16} + x^{15} + \cdots + x + 1)$.

Why it turns out like this, at present, I do not know. In order to find out, it will be necessary to make further detailed study, and this is surely a place where the profound beauty of algebra lies. I have a feeling that I have experienced a glimpse of this beauty.

[1], [16]	$0.9324722294 \pm 0.3612416662i$
[2], [15]	$0.7390089172 \pm 0.6736956436i$
[3], [14]	$0.4457383558 \pm 0.8951632914i$
[4], [13]	$0.0922683595 \pm 0.9957341763i$
[5], [12]	$-0.2736629901 \pm 0.9618256432i$
[6], [11]	$-0.6026346364 \pm 0.7980172273i$
[7], [10]	$-0.8502171357 \pm 0.5264321629i$
[8], [9]	$-0.9829730997 \pm 0.1837495178i$

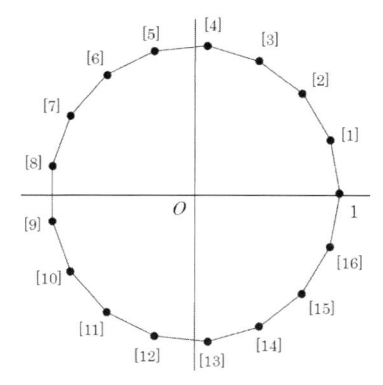

Figure 2: The Decomposition Process of the f-Step Cycle

Table 1. The 16 Roots

4 Constructing Roots, and More...

The value of $\cos\phi$ was expressed using roots, so let's look at their construction process specifically. Construction problems in mathematics are not solved by calculating numerical values using a computer and drawing diagrammatic illustrations. The figure must be drawn using only a ruler and compasses, and given two points on a plane (points 0 and 1, *i.e.*, a line segment) as a unit length. The ruler is only used for drawing straight lines and is considered incapable of measurement.

The problem is then how to draw the figure using root lengths. It is possible to construct the length of any root, in a manner similar to those shown in Figure 3. First, a unit square is drawn. By the triple angle theory, the diagonal edge has length $\sqrt{2}$. Secondly, a circular arc with radius $\sqrt{2}$ is drawn, and this length is transferred to the line extended from the base edge of the square. In this way, a rectangle with height 1 and width $\sqrt{2}$ can be constructed.

Once again applying the triple angle theory to this rectangle, the length of the diagonal is $\sqrt{3}$. By proceeding in this manner, the lengths $\sqrt{4}, \sqrt{5}, \sqrt{6}, \sqrt{7}$ and so on can be obtained. The construction of the length $\sqrt{17}$ can also be thus achieved.

Next, rather than the roots of integers, let's think about the construction of roots of arbitrary numerical values. Pages 50-51 of Kurata [2] discuss this in more detail and contain an explanation of how to construct fractions (of the form $\frac{a}{b}$) and roots (of the form $\sqrt{a}$). See Figures 4 and 5. Even in the present age, when computers have advanced so far, the construction method for the regular heptadecagon discovered by Gauss around 200 years ago in 1796 still impresses anew. For me it simply reaffirms Gauss' magnificent mental capacity and creative abilities.

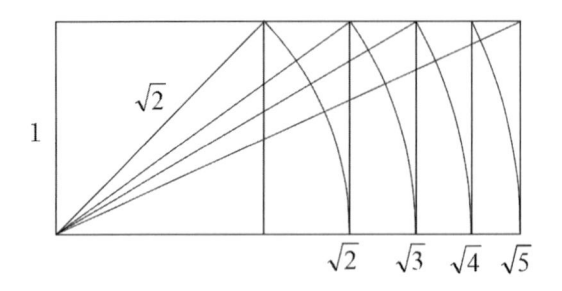

Figure 3: The Length of a Root

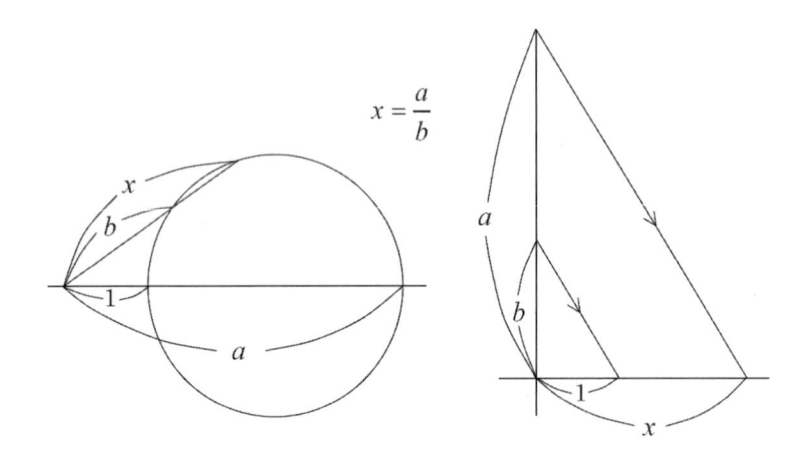

Figure 4: The Construction of a Fraction $(\frac{a}{b})$

References

[1] C.F. Gauss, (trans. by M. Takase), *Gauss Seisuron [Gauss' Number Theory, Disquisitiones Arithmeticae]*, Tokyo: Asakura, (1995).

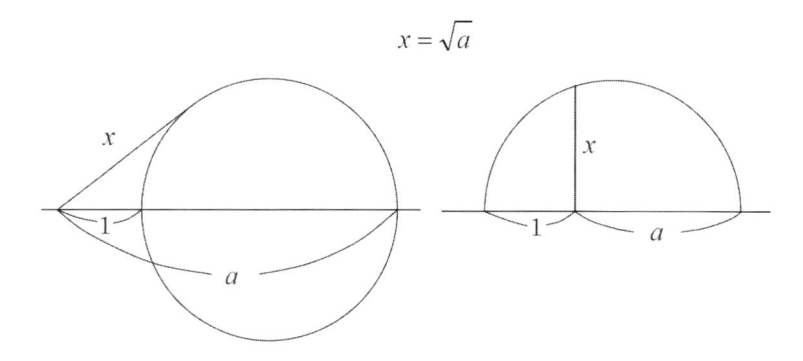

Figure 5: The Construction of a Root ($\sqrt{a}$)

[2] R. Kurata, *Gauss Enbun Hoteishikiron* [*Gauss' Theory of Cyclotonic Equations*], Nagoya: Kawai Cultural Education Research Laboratory, (1988).

[3] T. Takagi, *Kinsei Sugakushidan* [*History of Modern Mathematics*], 3rd edition, Tokyo: Kyoritsu, (1987).

CHAPTER 27
Sudoku: The New Smash Hit Puzzle Game

Abstract: *Sudoku* is now popular in many countries. This article explains the history and mathematics behind *Sudoku*. This includes the basic rules of *Sudoku*, a trial-and-error solution method, a discussion of the number of patterns completed, minimum *Sudoku* and the application of Euler's Latin squares. The popularity of *Sudoku* in the UK is also discussed.

AMS Subject Classification: 11A02, 00A09, 97A20
Key Words: *Sudoku*, Number-place, Minimum *Sudoku*, Latin squares

1 A *Sudoku* Book on the Best Seller List

In 2005 I had the chance to spend a year abroad conducting research at Cambridge University, and around May in 2005, one of my colleagues from England sent me the following inquiry: "Is there a general solution method for the type of Japanese puzzle known as *Sudoku*?" I tried to remember if I had heard of such a puzzle, and after thinking about it for a while I remembered that in the UK they pronounced it 'su doh ku' rather than 'suudoku.' Thinking about it some more, I remembered seeing some of the tired office workers on the train in Japan pitting themselves against this puzzle as if possessed. Some people regarded it as a slightly unhealthy puzzle hobby, and I had not once tried it. At that point I didn't even know the correct name for it.

Sudoku, written '数独' in Japanese, is the name used in the magazine *Pazuru Tsushin Nikori*, and is pronounced 'suu do ku.' *Sudoku* is a contraction of a phrase meaning 'numerals remaining alone,' and has been popular in Japan since around 1986. The puzzle had a previous incarnation in a puzzle magazine published in New York in the 1970s by the company Dell (no relation to the computer company of the present day). It was popular under the name 'number-place,' which refers to the placing of numerals, and subsequently became popular in Japan under the alternative name *Sudoku*. It is currently gaining in popularity in many countries.

I was rather delighted that it was not being called 'number-place,' but instead by its Japanese name, *Sudoku*. I investigated the extent to which people are familiar with this puzzle in the UK. As the result of British broadsheet newspapers such as *The Times* and *The Guardian* taking up *Sudoku* around the spring of 2005, all at once more general newspapers and community pages started publishing *Sudoku* problems as well. There are now books of problems that are published by these newspaper companies (see Figure 1). In Japanese terms, this corresponds to the *Yomiuri* or *Asahi* newspapers releasing books of *Sudoku* puzzles. If you went to a book store you would find a *Sudoku* book on the top of the best-sellers shelf. Once when I was watching a program on TV there was an interview with a celebrity who was asked "Do you enjoy *Sudoku*?" The reply was "Of course!"

(a)   (b)

Figure 1: The *Sudoku* Corner of the Shelves in a Bookshop (a). Standalone Book and Column in the *Guardian* (b).

The woman who lived next door to me enjoyed solving the *Sudoku* problem in her daily newspaper, and told me that it had become part of her daily routine. It seems as though people solving *Sudoku* had a certain glow.

One of the topics at a mathematics exhibition in July 2005 at a huge shopping center in Newcastle in the UK was a *Sudoku* corner. It was organized by the staff from a mathematical education research laboratory, who had prepared a complete set of *Sudoku* teaching materials. I was surprised that they had gone as far as producing educational resources. On the train back to Cambridge there were students earnestly wrestling with puzzle books. When I looked to see what puzzle they were working on, it was *Sudoku*. Recently there has also been *Sudoku* for game consoles, and I have seen people on buses enjoying *Sudoku* in such a way. I will touch upon it again below but *Sudoku*, unlike crossword puzzles, may be suited not to pencil and paper, but rather to personal computers.

Making it this far, *Sudoku* has really boomed. I made a pun in Japanese by joking that the 'doku' part means 'poison' since people who are addicted to it find they cannot escape, but I didn't manage to raise a laugh in the UK.

2 The Mathematics of *Sudoku*

There are many unexpected and hidden fans of *Sudoku*, but for the benefit of first-timers I'll explain the rules and a simple solution method.

The basic rules of *Sudoku*
Sudoku is one of those puzzles that involves filling in the numerals from 1 to 9 in a 9 × 9 square grid partitioned into 3 × 3 blocks. The thing is to place the numerals in such a way that no 3 × 3 block, nor any row nor column, contains a duplicated numeral. The derivation of the name *Sudoku* from 'numerals remaining alone' is well chosen, and refers to the fact that since duplications are forbidden, the numerals must remain alone. Of the 9 × 9 = 81 squares in the grid shown in Figure 2, 30 of them already have a number shown. These numbers are taken as hints and the remaining spaces are filled in. These hints may be given in a symmetrical or an asymmetrical fashion. The hints in Figure 2 are symmetrical [4].

5	3			7				
6			1	9	5			
	9	8					6	
8				6				3
4			8		3			1
7				2				6
	6					2	8	
			4	1	9			5
				8			7	9

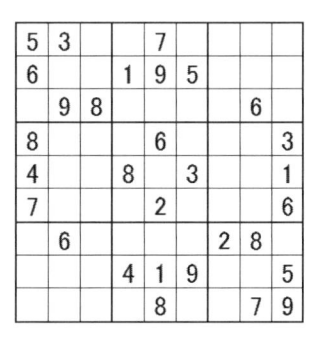

Figure 2: An Example *Sudoku*

Trial-and-error solution method

In the example, starting from the top, the first row already has a 5 in it so another 5 cannot be written in this row. The second row also already has a 5 in it, and the first column on the right-hand side already has a 5 in it so that neither this row nor column can take another 5. Elimination lines meaning that a 5 cannot be placed in these two rows or this column can therefore be drawn. Then when looking at the grid in terms of 3 × 3 blocks, the upper right block already contains a 6 so there is only one free box which can, and therefore must, contain a 5. The box in the 3rd row and 7th column is thus filled in (Figure 3).

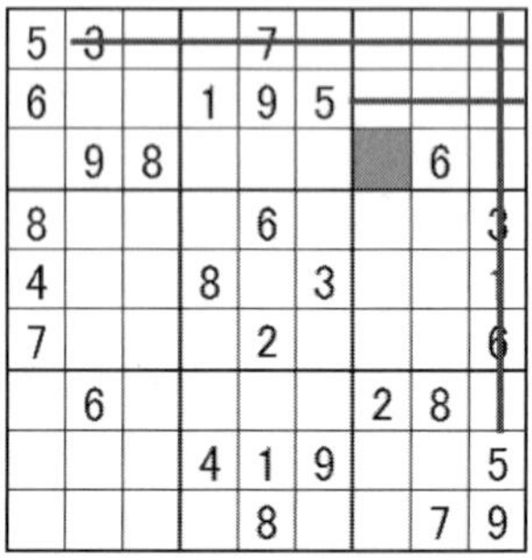

5	3			7				
6			1	9	5			
		9	8				6	
8				6				3
4			8		3			1
7				2				6
	6					2	8	
			4	1	9			5
				8			7	9

Figure 3: Solution Example Where the Position of a 5 Is Determined

By means of this process the numbers are filled in according to trial-and-error, but it's difficult to prove mathematically what kind of method is optimal. From experience, making use of the numbers filled in as hints and focusing on the places which already have many boxes filled in makes for a quick solution. When 8 of the 9 locations of a numeral are filled in, there is only one possible location left so its entry is fixed. When 7 locations are filled in, it is easier to find the answer than when 6 locations are filled in. From the hints in Figure 3, there are 5 locations containing an 8 and 2 locations containing a 2, so it is more efficient to start from the 8.

The number of patterns that can be generated
When all the numerals are filled in, the result looks like Figure 4. This is the completed pattern of the *Sudoku*. For each of the 9 rows, each of the 9 columns and each of the 3 × 3 blocks, there is no duplication of the numerals from 1 to 9.

Look at the completed figure in which the numerals are distributed evenly without any duplications. Don't you see a certain beauty here? How big is the number of such duplication-free distributions of the nu-

5	3	4	6	7	8	9	1	2
6	7	2	1	9	5	3	4	8
1	9	8	3	4	2	5	6	7
8	5	9	7	6	1	4	2	3
4	2	6	8	5	3	7	9	1
7	1	3	9	2	4	8	5	6
9	6	1	5	3	7	2	8	4
2	8	7	4	1	9	6	3	5
3	4	5	2	8	6	1	7	9

Figure 4: The Completed Pattern

merals? I supposed that it would probably be small, but investigating, I realized that a considerable number of patterns are possible. For example, for a small square grid of no more than 4×4 partitioned into 2×2 blocks, calculating the number of possible patterns revealed that there are 288. It is predictable that for 9×9 *Sudoku* problems, the number of patterns grows larger still. People have in fact calculated the number. Felgenhauer and Jarvis [2] stated it as

$$6,670,903,752,021,072,936,960 \approx 6.671 \times 10^{21}$$

according to the formula

$$9! \times 72^2 \times 2^7 \times 27,704,267,971.$$

The last term in the formula is a prime number. The fact that it is a prime means that it cannot be factorized, so this is the most refined form in which the numerical formula can be expressed. I have not verified for myself whether or not the value of this formula is correct, but it surely is. It means that a considerable number of patterns can be formed, and it seems unlikely that all the *Sudoku* problems will be exhausted.

Solution uniqueness

As I mentioned, *Sudoku* is a puzzle which involves writing numbers in the $9 \times 9 = 81$ squares of a grid. Some numbers are already shown as hints, and the number of hints is usually in the range of 20 to 36. In the case of Figure 4, there are 30 hints, which leads to the unique solution shown in Figure 4. If the number of hints is small, then the smaller it is, the more likely alternative solutions become. As extreme examples, if all of the $9 \times 9 = 81$ boxes are empty, then there is an almost unlimited number of solutions, but if 80 boxes are filled in then there is no scope for an alternative solution.

But what is the smallest number of hints that must be given in order to prevent alternative solutions? Although it does not constitute a proper mathematical proof, among those presented so far, the smallest has been an asymmetrical *Sudoku* with 17 hints. Among the symmetrical *Sudoku*, the smallest has 18 hints.

An example of a smallest 17-hint asymmetrical problem is shown in Figure 5. This was made public on the internet, and is from a set of 450 smallest *Sudoku* problems revealed by Royle [3]. There are few hints so it is difficult to solve, but the unique solution is fixed.

							3	1
	6				2			
					7			
		5		1		8		
	2						6	
			3				7	
					4		2	
		3		5				
7								

Figure 5: Example Smallest *Sudoku* (Asymmetrical, $n = 17$)

Application of Euler's Latin squares

The origin of *Sudoku* and number-place reach as far back as the Latin squares devised by the 18th-century Swiss mathematician Euler.

A Latin square has no duplicated numbers in its rows or its columns. In Figure 7, the numbers from 1 to 4 are filled-in in each row, and each column also has the numbers from 1 to 4, without duplications. A 4×4 Latin square is partitioned into 2×2 blocks. Adding the further condition that all the numbers from 1 to 4 must also be written in every block without duplications defines a *Sudoku*. *Sudoku* can therefore be regarded as an application of Latin squares.

In general, an $n^2 \times n^2$ Latin square can be divided into $n \times n$ blocks and made into a *Sudoku* problem. The 9×9 grid is popular for *Sudoku*, but at a higher level there are also 16×16 sized problems.

So *Sudoku* and Latin squares both involve many mathematical elements. They are not restricted to puzzles, and the following problems, which are convenient for university entrance examinations, can be con-

1	2	3	4
2	1	4	3
3	4	1	2
4	3	2	1

1	2	3	4
3	4	1	2
4	3	2	1
2	1	4	3

Latin square   Sudoku

Figure 6: Latin Squares and *Sudoku*

structed. With a 4×4 grid, how many Latin squares and how many *Sudoku* grids are possible? The answer is that there are $4! \times 3! \times 4 = 576$ Latin squares and $4! \times 2! \times 6 = 288$ *Sudoku* grids. This can be solved without a computer, using pencil and paper. Interested readers, please calculate the results!

Now, Euler demonstrated that the number of Latin squares with $n = 5$ is $5! \times 4! \times 56$ (1782). In relation to *Sudoku* grids, Bammel and Rothstein [1] demonstrated that for Latin squares with $n = 9$, the number of Latin squares is

$$9! \times 8! \times 377,597,570,964,258,816.$$

For the calculation, a computer of those times was used, namely, a PDP-10. Calculating the actual value of this formula yields

$$5,524,751,496,156,892,842,531,225,600$$

$$\approx 5.525 \times 10^{27}.$$

When $n = 9$, the number of Latin squares is of the order 10^{27} and the number of *Sudoku* grids is of the order 10^{21}, as shown above. About 10^{-6} of Latin squares are *Sudoku* grids, and they can be understood as further constrained patterns. Even so, the fact remains that there is an astronomical number of as many as 10^{21}.

For both Latin squares and *Sudoku* grids, the arrangement of numerals is truly balanced. Focusing on this well-balanced aspect, Latin squares already have an application in the field of experimental design methodology. Perhaps *Sudoku* arrangements will also find an application in statistics and open up a new field of research. They are promising for the future. By this token, I also attempted the challenge of *Sudoku* in England. I reached a level at which, no matter what the problem, I could solve it given enough time. *Sudoku* is certainly a type of puzzle that gives one a sense of achievement. Perhaps this is due to the quality of the numerals' balance. I also suspect that unlike crossword puzzles, they are not oriented towards the pencil and eraser style. This is because difficult problems involve many hypotheses regarding the arrangement of numerals, and upon reaching a dead-end it is not clear how far to back-track. Maybe these puzzles are suited to computers after all.

3   The Healthy Playful Psychology of the British

I thought that in Japan *Sudoku* was only popular among puzzle fans, but in the UK, the whole nation seems to enjoy it. I wonder why this is? I thought about where this difference might lie. By way of a similar kind of puzzle to *Sudoku*, the UK is the birthplace of crosswords. Amateur detective novels such as those of Agatha Christie and Conan Doyle were also born in the UK. It occurred to me that perhaps having plenty of rain and overcast weather, it might be a climate-related national characteristic that people discovered the pleasure of puzzles and mystery novels while spending much time indoors.

Besides the many people who are fond of *Sudoku* in Europe, including the UK, there are also many mathematicians researching *Sudoku*. Even as a research theme, *Sudoku* is not looked upon coldly, and many commendable pieces of research have appeared. In Japan, if one became absorbed in research on magic squares or other moth-eaten problems, these being among the interests of math fanatics, such researchers would probably be labeled as 'not mathematicians.'

The population of the UK is about 60 million, which is half that of Japan. From a biological perspective it would not be unreasonable for there to be twice as many excellent researchers in Japan. From the industrial revolution in the 18th century to the peak of the Victorian era, wasn't the UK's greatest achievement these mechanisms themselves? The historical process behind the success of the British Empire, by which much wealth was obtained from colonies in Asia and Africa, must not be forgotten, but there is an awareness that the country was developed though science and technology with mathematics as a prime example, and even now mathematics and mathematicians are held in high regard. Also, possessing the versatile language of English was an advantage. Japanese lacks versatility because it is difficult to achieve abundant literacy.

This generation in the UK turned out many scientists such as New-

Figure 7: The Main Entrance of Trinity College Where Newton and Ramanujan Both Studied

ton (differential calculus), Napier (logarithms), Boole (Boolean algebra), Cavendish (physics), Maxwell (electromagnetism), Faraday (chemistry) and Darwin (evolution). Even today, researchers gather from all over the world at the Cavendish Research Laboratory in Cambridge. Cambridge has educated 70 Nobel Prize winners, such as Watson and Crick who discovered the double helix of DNA. The mathematician Ramanujan was also enrolled in Trinity College at one point, and Wiles, who proved Fermat's last theorem, resided in Cambridge (see Figure 7.)

While they are aware of the pretension behind the thought that mathematics and science caused the country to flourish with prosperity, the people consider mathematics as important. In Japan on the other hand, as expressed by the words 'mathematics for examinations,' mathematics is only used as a method of sorting examinees by ability. I've presented a description somewhat biased towards the UK, but even removing this partiality, surely half of the description hits the mark.

I had the opportunity to study in Cambridge from April 2005. I graduated in mathematics in 1971, but had no postgraduate experience. Regarding research, most of my articles have been published in *Mathematics Seminar* and deal with mathematical games. From them I had 'The Flight Mechanism of Boomerangs,' 'The Puzzle of the Five Petals' and 'Constructions with Fixed Points' translated into English, and I then applied for overseas study. When I arrived in April, I was posted in the same relativity theory group as the astrophysicist Stephen Hawking and provided with a personal research room. Speaking in the extreme, it means to say that my article on boomerangs was acknowledged at Cambridge. I feel that the place has the fascination of an ancient university, which makes one want to try to see everything there is to see. It was a year during which I experienced the climate of the UK, where they accept *Sudoku* as legitimate mathematics, and I was compelled to think about the state of mathematics and mathematicians in Japan.

References

[1] S.E. Bammel and J. Rothstein, The Number of 9×9 Latin Squares, *Discrete Mathematics*, 11(1975), 93-95.

[2] B. Felgenhauer and F. Jarvis, Enumerating Possible *Sudoku* Grids, (2005). http://www.afjarvis.staff.shef.ac.uk/sudoku/sudoku.pdf.

[3] G. Royle, Minimum *Sudoku*, (2005).
http://school.maths.uwa.edu.au/~gordon/sudokumin.php.

[4] Sudoku article on Wikipedia, the free encyclopedia.
http://en.wikipedia.org/wiki/Sudoku

CHAPTER 28
Odd and Even Number Cultures

Abstract: Japanese prefer odd numbers, while Westerners prefer even numbers. This is clear from the distribution of number-related words in Japanese and English dictionaries. This paper explains the reason for this cultural difference by surveying the history of numbers, Yin-Yang thought from ancient China, ancient Greek philosophy and modern European mathematics. The author also mentions that while odd and even are only mathematical concepts, understanding the culture and history of individual countries contributes to world peace.

AMS Subject Classification: 01A27, 00A09, 97A40
Key Words: Odd number, Even number, Japanese and English, History of numbers, Yin-Yang thought

1 Japanese Prefer Odd Numbers

Japanese like the number "one." They also like the numbers "three," "five" and "seven," as in the Seven-Five-Three Festival (Shichigosan Festival) in which 3-year-old boys and girls, 5-year-old boys and 7-year-old girls go to shrines to celebrate their growth. It could have been ages six, four or two instead.similarly, there is a custom where festivals are held on odd-numbered days in odd-numbered months. January 1 (first day of the first month) is New Year's Day, March 3 (third day of the third month) the Girl's Festival, May 5 (fifth day of the fifth month) the Boy's Festival, July 7 (seventh day of the seventh month) the Star Festival and September 9 (ninth day of the ninth month) the Chrysanthemum Festival (this reinforces the significance of the odd-number nine). Haiku poetry is composed in lines of five-seven-five syllables; tanka poetry in lines of five-seven-five-seven-seven syllables; and Chinese poetry in quatrains with five-character (syllable) and seven-character lines. Cheer groups also clap their hands in three-three-seven beat rhythms.

295

In contrast, even numbers do not have good associations. "Two" means "to divide" (or "to part, separate"), "four" is associated with death and "six" as in the phrase *rokudenashi* means "good-for-nothing." At wedding ceremonies, people give gifts of 10,000, 30,000 or 50,000 yen. No one gives a gift of 20,000 or 40,000 yen. Similarly, at funerals the condolence payments are all in odd numbers. This may reflect the influence of Yin-Yang thought from China in which odd numbers are "Yang" (masculine). Hospital sickrooms and parking lots avoid the number "four" (which is homophonous with "death"). This is simply a matter of homophony and has no scientific grounds. Superstitions related to "four" are perhaps specific to Japan. People like odd numbers and dislike even numbers. However, there are two exceptions: "eight" (*hachi*) which can be taken as referring to "increasing prosperity" (because the Japanese character for eight is shaped like a fan, which starts narrow and broadens) and "nine" (*ku*) which can be taken as referring to "suffering," which is also read as *ku*.

2 The Distribution of Numerals

It is clear that a lot of odd numbers lie embedded and unseen in Japanese culture. Given this factor, I used a PC to conduct a survey of number-related words. In order to visualize the trend, I decided to search for word-initials (initial word-position agreement). In this way, I was able to search out all number-related words from one to nine.

In the case of English, the cardinal numbers from 1 to 9 are first of all one, two, three, four, five, six, seven, eight and nine, and the ordinal numbers are first, second, third, fourth, fifth, sixth, seventh, eighth and ninth. In addition, since there were more ordinal numbers, I added the following as data: single, double and triple corresponding to one, two and three; once, twice and thrice corresponding to one, two and three; half and quarter, referring, respectively to half of one and one quarter of one corresponding to two and four; and "couple" and "pair" referring to two.

As for the databases that were used for the acquisition of this data, the electronic version of the second edition of *Daijirin* (*Large Dictionary of Japanese*) published by Sanseido Co., Ltd. was used for the Japanese, and the sixth edition of *New English-Japanese Dictionary* published by Kenkyusha for the English. In addition, the free software supplied by Excite Japan Co., Ltd. (URL: http://www.excite.co.jp/) was used. Although the number of entries for English is slightly smaller than for

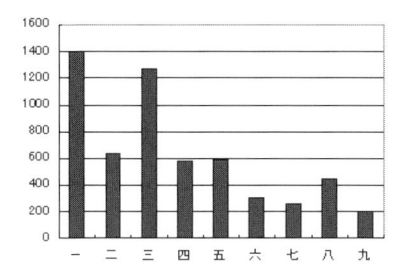

Figure 1: Distribution of Numerals in Japanese

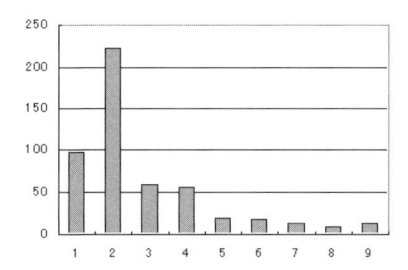

Figure 2: Distribution of Numerals in English

Japanese; the trend can be clearly seen (see Figures 1 and 2).

The graphs showing the distribution of numerals reveal the following points: numbers one and three are most common in Japanese, followed by two, four and five, which are used in approximately equal proportions, followed by six to nine in that order. Eight seems to be the most common in the six-to-nine sequence. In the case of English, two is most frequent, followed by one. Next, three and four are used in approximately equal proportions, and numbers five to nine trail without much difference between them. Looking at the distribution from a different angle, the odd numbers one, three and five seem to predominate in Japanese, while the even numbers two and four seem to predominate in English.

With that, it appears that the Japanese language has a cultural setting that favors the odd numbers three and five, while English has a cultural setting that favors the even numerals two, four and six. Language is a means of communication; as such it is a cultural legacy passed on through the long history of humanity. The human race is generally divided into white, black and yellow, and there are thought to be around 5,000 languages in the world. Besides English and Japanese, there must be frequency graphs of word distribution for various other languages.

3 Westerners Who Emphasize Symmetry

In Japan, there are always discussions about whether a two-party system can be realized. It is thought that politics will improve if a two-party system such as the American Democratic and Republican parties and the British Labor and Conservative parties can be realized.

Since the Meiji Era, however, Japan has not realized a two-party system. Currently, The Democratic Party of Japan is aiming at a two-party system, but it seems that Japan is not congenial to such a system. Is it too bold to suggest that this is because Japanese culture is conducive to odd numbers, while American and British culture is conducive to even numbers? For Japan, it is either one or three.

Moreover, a new 2000 yen note was issued in 2000, but the use of it hasn't taken root. One of the causes is thought to be the fact that current vending machines do not support this new note, but I suspect there may be other causes. America has 2-dollar and 20-dollar notes, and the UK has the 2-pound coin and 20-pound note. Why is this? Doesn't this suggest that there may be some connection with cultures that are conducive to odd numbers and those that are conducive to even numbers?

There is an interesting story about the rainbow. According to a book by Takao Suzuki entitled *Nihongo to Gaikokugo* (*Japanese and Foreign Languages*) published by Iwanami Shinsho, the rainbow is perceived to have seven colors in Japanese culture. He presents a very interesting analysis of why the rainbow is perceived as having five or six colors, depending on the country. In Japan, the seven primary colors of the rainbow are 赤 (*aka*), 橙 (*dai-dai*), 黄 (*ki*), 緑 (*midori*), 青 (*ao*), 藍 (*ai*), and 紫 (*murasaki*); these correspond to red, orange, yellow, green, blue, indigo and violet in English. When some Westerners refer to the rainbow as having only six colors, indigo is omitted. However, a rainbow actually consists of a continuous spectrum of colors, so neither perception is correct. Therefore, isn't it possible that people's perception of the rainbow as consisting of either seven or six colors reflects whether their culture favors odd numbers or even numbers?

The *gobosei* (pentagram) and *rokubosei* (hexagram) used in Chinese/Japanese-style fortune-telling represent shapes of stars. The *gobosei*, also known as the pentagram, is well known as forming the five-petaled bellflower family crest of Abe Seimei (a noted fortune-teller in Japan). The five points of the *gobosei* (pentagram) are represented by wood, fire, earth, metal and water in the Yin-Yang five elements of thinking. The *rokubosei* (hexagram), which is also the shape of the "Star of David," the symbol of Judaism, is written with two triangles, and is also regarded as symbolizing the union of Yin (female) and Yang (male). Although stars are actually spherical in shape, they are expressed by the shapes of the pentagon and hexagon. Hence, the *gobosei* (pentagram) can also be regarded as reflecting a worship of odd numbers while the *rokugosei* (hexagram) may be regarded as reflecting a worship of even numbers.

Next I will cite examples of the way in which numbers are perceived differently depending on the culture of the country. There is a Japanese saying that goes *gojuppo hyakupo* (50 steps or 100 steps). The *Modern English-Japanese dictionary* (Kenkyusha, 1973) translates this saying as "It's six of one and half-a-dozen of the other." When there is little difference between two things, this is referred to as "50 or 100" in Japanese and "six or a half-dozen" in English. The duodecimal and sexagesimal systems spread in the West, and these are rational numbering systems with many divisors. Is the history of odd and even numbers in the West and the East also reflected in proverbs?

There is a proverb that goes *sanninyoreba monjunochie* (the counsel of three persons will result in wisdom like that of Monju, the Bodhisattva of Wisdom). These proverbs purports that even three fools can come together and have discussions that will result in wisdom like that of Monju. According to the *New English-Japanese Dictionary* (Kenkyusha, 1995), the corresponding English proverb is "Two heads are better than one." In Japanese, three are better than two, while in English two are better than one. Doesn't this perhaps show the difference between favoring odd numbers (three people) and favoring even numbers (two people)?

4 Odd and Even Numbers in Various Languages

What happens to odd and even numbers in other languages? In ancient China, the characters 奇 (*qi* or *ki*) and Char1 (*ou* or *gu*) were used to express odd and even; however, nowadays 奇 (*qi* or *ki*) and 偶 (*ou* or *gu*) are used. The character Char1 means "ploughshare" (two persons lined up to plough together). Similarly, singular numbers are sometimes used to refer to odd numbers and dual numbers as even numbers. Korean Hangul writing uses Char2 to refer to the odd number and Char3, to the even number. In Indonesian, the term *ganjil* refers to the odd number and the term *genap*, to the even number. Like so, China has had a great influence in the East.

Char1: 耦   Char2: 기수   Char3: 우수

What about the West? English uses the terms "even number" and "odd number." In Spanish, the terms used are *número par* (even number) and *número impar* (odd number). In French, the terms used are *nombre pair* (even number) and *nombre impair* (odd number). In Italian, these are *nùmero pari* (even number) and *nùmero dispari* (odd number). In German, they are *gerade Zahl* (even number) and *ungerade Zahl* (odd number). In European languages, the word referring to the even number is the basic term and the term referring to the odd number is the negated form of the basic term. In old English, the term for the odd number was also referred to as the "uneven number," following the same principle. In current English, the completely different term "odd number" is used.

By now, one wonders why Japanese prefer odd numbers. What could be the reason for liking numbers known as "odd," which means eccentric, a crackpot and a strange person? A preference for odd numbers is in itself strange.

Let's look at words that use the kanji 奇 (*qi* or *ki*.) There are words with positive connotations, such as 奇才 (*kisai*, a prodigy), 奇抜 (*kibatsu*, unique), 奇特 (*kitoku*, praiseworthy) and 奇跡的 (*kisekiteki*, miraculous). There are also words with negative connotations, such as 奇怪 (*kikai*, mysterious), 奇行 (*kiko*, eccentric conduct), 奇声 (*kisei*, strange voice) and 奇人 (*kijin*, an eccentric person). The ratio is about 50:50. What is odd, however, is that the positive connotations associated with 奇 (*qi* or *ki*) are disappearing and only the negative connotations seem to be emphasized. In English, the term "odd" in "odd number" has meanings such as strange, queer, peculiar and one-sided. These have mostly negative connotations.

Now, when did the connection occur between the Japanese and English language in which 奇数 (*kisu*) became to be referred to as the odd number and 偶数 (*gusu*) as the even number? Most Japanese mathematical terms were defined by The Physical Society of Japan, which was established in 1877 during the Meiji Era. The definitions in which 奇数 (*kisu*) became to be referred to as the odd number and 偶数 (*gusu*) as the even number were officially established in 1881.

Then, what was the case before the Meiji Era? Were these terms for even number and odd number translated by the English or the Japanese? Or were there separate original terms for them? It is necessary to look into the history of mathematics and the history of numbers in particular to find out why.

5 Odd Numbers Were Regarded as "Good" in Ancient Greece

The history of numbers begins with natural numbers, evolving to negative numbers, integers, fractions, decimals, rational numbers, irrational numbers, Gauss complex numbers, transcendental numbers and Kline quaternions. What is discussed here are the natural numbers, namely the ten initial numbers, one to ten.

The aim is to determine when the terms "even number" and "odd number" came to be used. According to Denis Guedj in *L'empire des nombres* (Sogensha Inc.), the Pythagoreans of the 5th and 4th centuries B.C. were the first to distinguish odd numbers from even ones.

Their use of this concept paved the way for a variety of results. The line of natural numbers is made up of the infinite repetition of even and odd numbers. Since the prime numbers were found in order to classify natural numbers, the classification of even numbers and odd numbers was therefore conducted by the initial prime number "two."

Once the numbers were classified into the two categories, the "odd numbers" and the "even numbers," which set did the ancient Greeks prefer? Wataru Uegaki makes the following point in *The Origin of Greek Mathematics* (Nippon Hyoronsha Co., Ltd.):Pythagoras not only linked everything to numbers, but he tried to express everything in numbers. First, Pythagoras classified numbers into odd numbers and even numbers. Then, since odd numbers cannot be divided at will, and with the belief that what cannot be broken down is 'complete,' he connected odd numbers to "completeness," "mysteriousness" and "finiteness." In contrast, even numbers, because they could be divided into two, were given the opposite attributes.

Aristotle summarized his views on the matter in Vol. 1, Chapter 5 of *Metaphysics* (Iwanami Bunko) as follows. He said there are ten principles, which he listed in a table of oppositions. In other words, there are finite and infinite, odd and even, one and many, right and left, male and female, stillness and motion, straight and crooked, light and darkness, good and evil, squares and rectangles. This list of oppositions is known as the table of categories of the Pythagoras school or the table of contrary concepts. In general, he considered the first-listed concepts to be 'good' and with 'form,' while he considered the second-listed concepts to be 'bad' and 'material.'

6 Odd Numbers and Even Numbers in the Yin-Yang Thought

In the times of the ancient Greeks, odd numbers were regarded as 'good.' This resembles the situation in Yin-Yang thought in China, discussed below. Was Pythagoras of ancient Greece the first to divide numbers into odd and even? I suspected that the ancient Chinese may have been the first. Thus, I looked into the fortune-telling of the Yi Ching, associated with Yin-Yang thought in China. Yi Ching thought-based fortune-telling originated in the Zhou Dynasty of ancient China, which lasted from the 12th to 3rd century B.C. This is roughly the same time period as that of ancient Greece. It is not certain which was first, but there are certain close resemblances between their thought, and therefore one can presume that there was some sort of reciprocal influence between these two centers of thought.

Now, let me briefly describe Yi Ching fortune-telling or basically, Yin-Yang thought. The core is Yin and Yang (the alternation of masculine and feminine). Originally, there were no *kanji* (Chinese characters) for Yin and Yang. Previously, there was 'softness' and 'hardness' ('weak' and 'strong'.) And before that, there were other designations and characters, but I will not delve into that matter here. Yi Ching thought assigned the principles of movement-stillness and hardness-softness to odd and even numbers. Yang was strong and Yin was compliant. Yang was movement and Yin was stillness. All things of the natural world and the world of humans were arranged under the two Yin-Yang principles. The things that were Yang were the heavens, the sun, fathers, males, benevolence, top, front, brightness, going, midday, respect, nobility and good fortune, while the things that belonged to Yin were the earth, the moon, mothers, females, justice, below, behind, darkness, coming, night, lowliness, humbleness and misfortune.

Thus, although Yi Ching thought is based on the concept of two alternating opposites, Yin and Yang, the next point, which is written on page 40 of the Iwanami Bunko edition of *Yi Ching* (Book 1), is very important. A male is Yang in relation to a female; however, that same male is Yin in relation to his parents. A female is Yin, but is Yang in relation to her child. Front is Yang in relation to behind, but is Yin in relation to things that are in front of it. Yin and Yang imply infinite change; this is the message imparted by Yi Ching thought. Hence, Yin and Yang are not universal absolutes. It is emphasized that Yin and Yang are simply polar opposites and not a case of one being superior or inferior to the other. In contrast, it is only in modern times that Yang became to be considered good and superior, and Yin as bad and inferior.

7 Contributing to Peace through an Understanding of Numbers

In Yin-Yang thought, numbers are divided into Yin and Yang. The "Transmitted Interpretations of the Zhou-yi" in *Yi-Ching* (Book 2, Iwanami Bunko) arranges the numbers from one to ten in the following, alternating a Heaven and Earth pattern: "Heaven is one, Earth is two; Heaven is three, Earth is four; Heaven is five, Earth is six; Heaven is seven, Earth is eight; Heaven is nine, Earth is ten; Heaven accounts for five numbers and Earth five." The terminology may vary between Heaven and Earth, Yin and Yang, Hard and Soft and so on, but the numbers are just arranged in a pattern and do not seem to be conceptual like the odd and even numbering of today.

The emergence of the concept of odd and even really had to wait on the development of mathematics. The following interesting example is found in the *Sunzi San Jing*, an ancient Chinese book on arithmetic. There is a pregnant woman who is 29 years old, and the month in which the baby is due is September. The question is whether the baby will be male or female. The answer is male, for the following reason. First of all, the number forty-nine is posited. The month of delivery is added, and the age of the mother is subtracted. Then, one for the Heaven, two for the Earth, three for man, four for the four seasons, five for the five elements, six for the six tones, seven for the seven stars, eight for the winds of the eight directions and nine for the nine states are subtracted. If the remaining number is an odd number, the baby will be male; if it is even, the baby will be female. If one tries this calculation, the answer will not agree. However, it is obvious that the concept of odd and even existed.

The development of the mathematics of ancient Greece, which developed under the influence of Egypt and Babylonia, stopped temporarily, but then led to the development of the mathematics of the Arab world and India. Arabic algebra was transmitted to Europe via Italy during the Crusades. However, the establishment of modern mathematics was still well in the future. Modern mathematics, as represented by Newton, valued rationality and science. Consequently, it seems to have abandoned the ideas from ancient Chinese Yin-Yang thought and ancient Greek philosophies, in which the odd number was male and the even number female. When counting numbers, odd numbers were incomplete, in-between numbers, while even numbers were certainly the more rational.

In modern times, when the mathematics of the East and that of the West, which developed independently, merged, historical and cultural differences between the terms were probably exposed when the mathematical terminology claimed *kisu* (奇数) to be odd numbers and *gusu* (偶数) to be even numbers. In contrast to the East, where odd numbers were positive and good, in the West, they were incomplete and superfluous.

I have argued that the East has a culture of odd numbers and the West a culture of even numbers. However, what about the Middle East? That would be interesting to look into. Moreover, the sun is positive and good in both the East and the West, whereas in the Middle East, the moon has positive connotations, which is also different from the Yin-Yang meaning in the East. Moreover, Arabic writing is just the opposite of English; teletext on news broadcast by the Middle Eastern satellite broadcasting station AL Jazeera moves from left to right.

Odd numbers and even numbers, the sun and the moon, the direction of writing and so on are opposing concepts; however, opposites are not eternally opposite. In dialectics, oppositions are contradictions; however, they are unified by sublation and the combination of thesis and antithesis. When discussing the Iraq issue and other international disputes, there are surely matters that cannot be discussed only in political and economic terms. Differences in historical relationships lie deep in individual cultures and histories. I believe that it is possible to find ways leading to peaceful settlements by mutually acknowledging such differences. Although the odd number and even number are mathematical concepts, they seem to be deeply rooted in politics and society as well [1].

Reference

[1] Y. Nishiyama, Su no Bunkashi [The Cultural History of Numbers], *Keizaishi Kenkyu* [*Studies in Economic History*], 8(2004), 146-174, (2004).

CHAPTER **29**
Cultures of Curves and Straight Lines

Abstract: Japanese buildings are constructed with straight lines. European buildings, however, tend to have arched rather than straight horizontal lines. Roofs, window frames and the tops of doors are often arched. European buildings are made not from wood but from stone. The reason for these differences is explained by mechanical theory.

AMS Subject Classification: 51A02, 00A09, 97A20
Key Words: Earthquake, Architecture of window frame, Curves and straight lines, Culture of stone architecture

1 Cultures Building with Stone and Wood

It was on my first trip to France in 2000 that I felt that there was something different between European and Japanese buildings. On my short trip of 5 days in Paris and 5 days in Avignon in Southern France, whenever I looked around me, all the doors and windows were arched, which surprised me a little. I have lived in Japan for many years, so I have come to believe that buildings are constructed with straight lines. Pillars stand up vertically from the ground. Beams span the gap between pillars horizontally. Both pillars and beams are straight. Doors and window frames are rectangular, and so are also made up of straight lines. Roofs are sloped but they are roughly straight. The verticals and horizontals used in reinforced concrete buildings are both straight.

European buildings, however, are not always horizontally straight. They use arches, that is to say, curved lines. Roofs are round and domed. Window frames and the tops of doors are arched and round, as are bridges. The vertical direction is straight, but horizontally, there are few straight lines to be seen. The arched European buildings that I saw in the paintings at the Louvre were beautiful.

So why are European buildings curved rather than straight? I wondered if perhaps Europeans simply preferred curved lines and other such things, but the question was answered when I went to Avignon in the South of France.

Near Avignon there are small towns such as Arles and Orange, and here there are the remains of buildings constructed during the time of ancient Rome. The outer surfaces of these remains have fallen away, so the architectural structures and construction processes can be easily understood. The reason that the tops of the door and window frames are arched is because these buildings were constructed of stone. This simple fact did not occur to me while I was in Japan. European buildings are made of stone, while Japanese buildings are made of wood or reinforced concrete. When building with stone, the upper parts of door and window frames are rounded, but when using wood, they are straight. The architectural materials therefore determine the form of buildings.

2 The Climate and Architecture of the United Kingdom

In 2004 I participated in the International Congress on Mathematical Education (ICME10) in Denmark, and in 2005 I had the opportunity to spend a year doing research in the United Kingdom. At that time, as an amateur, I investigated the forms of European architecture. The characteristics of the windows are as follows. The first thing I noticed was that the proportion of the vertical and horizontal dimensions of windows was larger than in Japan, *i.e.*, they are taller than in Japan. There were even windows twice as tall as they were wide. Also, in Japan, windows may be composed of two panes which can be moved horizontally, but in the UK it is common for windows to have only one pane, which is frequently fixed. Although they sometimes move, it is only by a really small distance. I did wonder whether the ventilation was satisfactory, but it seems that the spaces are well ventilated.

Houses in the UK typically have full central heating systems, and in such cases there is no need for free-standing gas heaters or electric blankets that we generally need to use in Japan. Since there were no small gas heaters, the volume of carbon dioxide produced was also small, and perhaps there wasn't so much need for ventilation. Tokyo is at latitude 36° and London at 54°. Sapporo is at 43°, so London is situated considerably further north than Hokkaido but the warm current from the Gulf of Mexico somehow tempers the coldness. As they say, the weather in the UK has "four seasons in a single day," even if it is warm in the day and the sun is scorching, it can suddenly start raining or become chilly.

The reason the weather changes so drastically is that the landscape is composed of rolling hills without mountains. There are mountains in Scotland to the north, but in England in the south, there aren't any. In Japan the mountains obstruct the flow of the air, but this doesn't happen in the UK. It is reasonable to regard the UK as being under the influence of a marine climate. According to travel books, the time difference between the UK and Japan is 8 or 9 hours depending on daylight savings. Daylight saving time is not well known in Japan, but it is adopted in the high latitude countries of Northern Europe. Countries with a high latitude have long days in summer and long nights in the winter. In summer, even after 9 pm it may be bright and people continue to enjoy socializing outside. But in the winter, it may be dark by 3 pm and cold. If you visit the UK to study, it's probably a good idea to go for the six months between April and September, and better to move to Italy or similar for the remaining half of the year.

I had a rather interesting experience with the daylight saving time system. I took a short trip from the UK to Belgium for 3 days and 2 nights. After 2 hours and 40 minutes on the European equivalent of the *Shinkansen* (bullet train), I reached the capital of Belgium, Brussels. I left on the 29th of October and returned on the 31st. The time difference between the UK and Belgium is 1 hour. While away, I set my watch to the local time. I was used to doing this, but October 30 (a Sunday) was the day on which the daylight saving time changeover occurred and I mixed myself up. Among my mobile phone, wrist watch, the timer on my computer, the wall clock in my lodgings, the clock on the TV and the clocks in the street, some were changed and others not, and I was confused. When I went for lunch at the college dining hall, there was no one there. I felt bewildered that whole day.

The highest temperature in the summer is around 25°C, and even considering such days as hot, they never seem to continue for more than 5 days in a row. Since enduring just 5 or so consecutive days of heat is sufficient, they don't have air conditioners in the UK but instead make do with fans. During my stay, I hardly even used my fan. While I was visiting Cambridge, I visited London at the weekends to see galleries and museums.

London is only about 1 hour from Cambridge by train, and during my trips it felt like the diesel locomotives were sliding on water. If you're wondering why, it's because, unlike Japan, there's no need to worry about the rails expanding in the heat, so it's not necessary to leave big gaps between the links. The gaps between the rails in the UK are probably small, and I hardly heard the sound of the rails clattering.

The UK only very rarely has noticeable earthquakes, and there is no record in history of an earthquake measuring above 5 on the Richter scale. The surface of the Earth is divided into tens of tectonic plates. Each plate may be 100 km thick, and they are all moving in their own directions. Minor earthquakes occur at the oceanic ridge where plates are formed, and major ones occur where plates collide, like in the Himalayas, and where plates subduct, or sink down, like in the Japan trench. There are 4 plates in the vicinity of Japan. In Eastern Japan the Pacific plate subducts beneath the North American plate, and in Western Japan the Philippine sea plate subducts beneath the Eurasian one. Also, from the Izu Peninsula to the Bonin Islands the Pacific plate subducts beneath the Philippine sea plate. When these plates push against each other and exceed the limits of resistance, an earthquake occurs. Since there is a gathering of plates in this way, Japan experiences many earthquakes.

The UK is within the Eurasian plate, and since it is not on a boundary between plates, no large earthquakes occur. King's College in Cambridge was constructed in 1446, so it has over 500 years of history. They say that no earthquakes have occurred during this long period of time. The buildings don't even have reinforced frames. They are just piles of stone, and they do a good job of standing up considering that they don't have a reinforced structure. If there were such tall stone buildings in Japan they would be unlikely to survive because of the earthquakes.

3 The 'This and That' of Window Shapes

Detached houses and multi-dwelling flats are representative types of residence in the UK. Houses tend to have many rooms and a garden, and one must be of some means to live in one. Flats correspond to Japanese *mansions* (which are apartments made of brick or stone, and not the huge stately homes known as mansion houses in Europe!) although they are usually bigger than in Japan, and the UK also has what are known as *sangaitate* (three-storey flats) in Japan. The ground floor, first floor and second floor in the UK correspond to the first, second and third floors in Japan. Houses and flats are mostly constructed by stacking up bricks of a uniform nature. In Cambridge I sometimes saw buildings in mid-construction, and while they were attaching bricks with cement, they never used any reinforcement. The bricks are just layered up. I wondered whether they were safe, but for a country with no earthquakes, they surely are.

I took some photographs of windows in Cambridge city center. Figure 1 shows a relatively new window frame in a flat. The glass part is rectangular and composed of straight lines, and the bricks are layered in a radial pattern. The bottom part is cut by a straight line. Perhaps stone buildings evolved, and naturally ended up in this current form. The top part of this form of window captured my attention completely. The bricks are arranged prettily and stacked up. Only the top part is layered differently, marring the aesthetic appearance. Figure 2 shows a window frame from a building that I believe is a little older. The glass part is straight and rectangular, but the bricks in the upper part are not straight. Although it is less than 10 cm tall, there is an arch. I named this the "ultimate window frame."

Figure 1: Representative Window Frame from a Private Dwelling

Figure 2: No Matter How Slight, There Is an Arch (Ultimate Window Frame)

However, one may also sometimes find window frames like that shown in Figure 3. This incorporates a mechanically impossible method of layering the blocks. In order to hold the blocks horizontally, a force of unlimited strength is required at both ends. If this isn't the case, then the strength of the glass window on the inside must be holding up the blocks and preventing them from falling down.

If you are wondering if all window frames are arched, this is not the case. There were also some constructed with straight lines like that shown in Figure 4. The top part of the opening is horizontal and straight, but the blocks at each end are stacked in slopes. Diagonal arrangements of blocks are dynamically possible, but the blocks along the top are still a concern. Perhaps the glass inside or a metal frame supports the blocks above so that they don't fall down.

Figure 5 shows an arched gateway. Pay attention to the way it is stacked. The stones are stacked in two radial layers. Using an arch rather than a horizontal pattern means that the blocks do not fall down.

Figure 6 shows a window from the Fitzwilliam Museum. Like the

Figure 3: A Mechanically Impossible Window

Figure 4: Polygonal Window

gateway shown in Figure 5, the marble blocks are stacked in three radial layers. As a consequence, the window is also arched (in a circle). When stacked in an arch, the top center piece provides great strength, so large pieces of marble were used. The center block is in the exact middle so it is angled vertically at 90°. Blocks cannot be stacked at 90°, so a large piece of marble is probably used in order to give it a slope. The fact that the block in the center is the number one problem when building an arch is considered the fate of stone architecture.

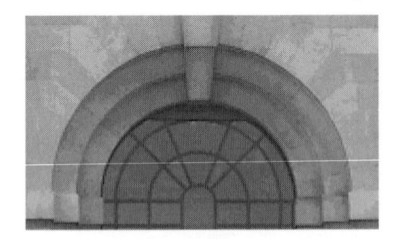

Figure 5: Arched Gateway

Figure 6: Fitzwilliam Museum (Cambridge)

About 1 hour away from Cambridge by train is the city of Peterborough, which has a Gothic cathedral. Figure 7 shows the window frames from a Gothic church there. Europe has a Christian culture and all the countries have churches and cathedrals. The form of the window frames is not circular like that shown in Figure 6, it is pointed. This is a basic form of Gothic architecture, and allows for the construction of tall buildings. I was interested in this form, and wondered whether it might be a harmonic progression viewed in terms of building blocks, *i.e.*, whether it takes the form of a Log function. However, investigating it by tracing over a photograph, I discovered that it is no more than a pair of circular arcs in composition. Instead of using one circular arc with an internal angle of 180°, using two circular arcs with internal angles of 60° allows the arch to be completed with a reduced burden on the central part. Functionally speaking, a straight horizontal form is known to be the best, but it is impossible to stack stones horizontally, so this dilemma means that they must be stacked in a circles or circular arcs.

Figure 8 shows a theater gateway that I found by chance when visiting Edinburgh in Scotland. Wouldn't the key stone in the center of the gateway normally spoil the aesthetics? No doubt this is why the architect made the key stone with a lion's tongue design in order to hide it.

Figure 7: A Gothic Window (Peterborough)

Figure 8: The Art of Hiding the Keystone (Edinburgh)

Figure 9 shows a photograph of the doorway to the flat where I was living. The door is rectangular, with straight vertical and horizontal lines. There is no need to make the glass with a circular arc, but for some reason it is made with an arch. Perhaps this is due to the influence of the culture of stone architecture in Europe.

Figure 9: An Unnecessary Arch in a Door (Flat Lodgings)

4 Stone Buildings as Assets

All bodies on the surface of the Earth are under the influence of its attractive force. We know just how weak the buildings under the influence of gravity are in a horizontal direction. For example, metal electricity pylons and the wires stretched between them sag under the force of gravity. In order to stretch them out straight, an infinitely large force would be required. This is impossible, so they take the form of catenary curves. Gravity acts in the same way on the beams of buildings, and the deformation due to flexion means that it is difficult for them to be kept horizontal.

For the stones in the top parts of window frames, the weight of the stones themselves and the material above acts to produce a 'bending' force encouraging the central part to droop downwards. The relationship between the stress and distortion when wood or metal is stretched and deformed is expressed as a stress-distortion curve. The initial deformation occurs as an elastic deformation, and the stress-distortion curve has a linear relationship. Increased stress gives rise to plastic deformations and the relationship deviates from a straight line. Further increasing beyond the yielding point causes the material to break.

Wood is pliable and has elasticity. Relatively long pieces are used as material for posts and beams. It is easy to work with, but being easy to burn it has the drawback of being susceptible to fire. Metal is stronger and more elastic than wood, but may be prone to rust. A material combining the properties of metal and stone is reinforced concrete.

Stone is hard, strong, heavy and not an elastic body. It is difficult to work with, and it is hard to obtain long thin pieces of stone like granite, marble and limestone to use as a material horizontally. Granite is used for the outer parts of buildings, while marble and limestone are used for interior decoration. Limestone is a rock composed mostly of calcium carbonate, which is formed from the carbonated lime in the fossils of living creatures with shells and constituents in the sea that fell to the bottom as sediment. The influence of heat on this limestone causes it to transform and re-crystallize as marble. Limestone and marble are both susceptible to acid rain.

I think I understand why European nations strongly desire to enact the Kyoto Protocol to prevent global warming which regulates carbon dioxide emissions. They are extremely opposed to reinforced concrete buildings. Metal rusts and has a short lifespan, but stone buildings are known to last for 400 to 500 years. While I was studying abroad, the college staff were debating over the construction of a new school building, and hearing that they had no intention of considering using reinforced concrete made a very strong impression on me.

CHAPTER 30
Counting with the Fingers

Abstract: It has been written that across the world there are 27 types of counting method using the fingers. For example, there are various methods for starting to count '1' using either the index finger, thumb or little finger. In ancient Rome, people counted '1' by bending their left little finger. Counting using the fingers differs according to region, ethnicity and historical period. This chapter also discusses the 'evolutionary theory' of counting methods.

AMS Subject Classification: 01A02, 00A09, 97A20
Key Words: Finger counting, Counting with the fingers

1 French People Start with the Thumb

It was the summer of 2000. On my first family trip to Europe I had something of a cross-cultural experience. After visiting the Sacre Coeur Basilica on the Montmartre butte in the center of Paris I was hungry, and spotting an open-air restaurant, something that looked like a crepe sprinkled with sugar caught my eye. At the restaurant I said, 'One of these please' in English but they didn't understand me. So then I said 'this,' pointed at the food, and held up my index finger to indicate '1.' Once again they didn't understand. After another two or three attempts in silent dialog, I realized that while Japanese people indicate 1 using the index finger turned away toward the other person, French people use their thumb. Now that I mention it, I remember it was like a hitchhiker using their thumb to flag down a car, which felt a bit strange. Why is it that people who are born on the same Earth use different counting methods according to their country or region? This question has not left my head ever since.

1   2   3

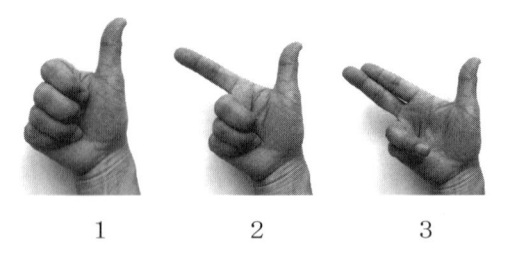

Figure 1: Counting with the Thumb (Germany and France)

2 Is Counting with Bent Fingers Inherent to Japan?

When I was working as a member of the international exchange committee at university, I talked about this issue with a Swedish researcher, who told me that they definitely start counting from the thumb in Northern Europe. What's more, Europeans begin counting with the hand closed, *i.e.*, with the 'scissors, paper, stone' symbol for 'stone,' while Japanese people start from an open hand, *i.e.*, with the symbol for 'paper.' We Japanese have a tradition of beginning to count with 'paper,' and counting "one, two, three..." by bending down first the thumb and then each finger one after another. This goes as far as having the phrase *yubiori kazoeru* (finger-folding counting) in Japanese, so one can believe that this counting method has been used in Japan for a very long time. The Swedish researcher suggested that since Sweden, being in Northern Europe, is cold, people always keep their hands closed, while Japan is a warm country so people always keep their hands open. This means that the weather and climate is related to the way people count. This is an interesting idea. I wonder if this is really true.

This counting method also has a touch of a sad story associated with it. The story happened in India during the Second World War. An Indian girl had to introduce an Englishman who had called at her home to some Asian female friends. The problem was that the women were Japanese, and if this were known, they would be arrested in an instant. So she concealed their nationality and said that they were Chinese. The Englishman harbored some doubt, and he called upon them to 'try counting with your fingers, up to 5.' They say the Englishman saw through the fact that they were Japanese by the way they counted? starting with their hands open, and bending their thumb and fingers in one by one [1].

       1   2   3

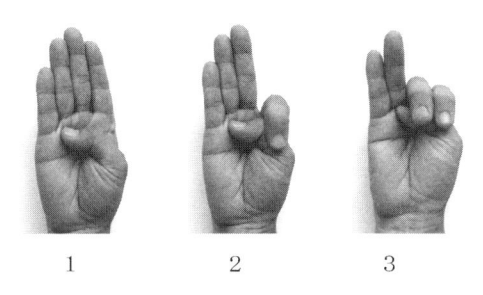

Figure 2: Bending the Fingers Beginning with the Thumb (Japan)

3 Chinese People Who Count to 10 on One Hand

I once visited a university in China regarding an exchange program. Mentioning the topic of counting methods at a social reception, I was told that, "Chinese people and Japanese people count in the same way from the index finger, but we count to 10 on one hand," and I was taught this counting method. 1 to 5 are counted in the same way as in Japan, but 6 is made by extending the thumb and little finger from the 'stone' position. It looks like the gesture for a telephone. In Hawaii people say "Hang loose!" and make this gesture as a greeting. Apparently this means "Take it easy." Perhaps in the Chinese gesture, the thumb represents 5 and the little finger represents 1, and these combine to make 6.

7 is represented by extending the triple of the thumb, the index finger and the middle finger. Perhaps the thumb represents 5, and the other fingers combine to make the remaining 2. 8 is represented by using the thumb and index finger, and opening them to form the Chinese character for 8, '八.' 9 is represented by making the shape of the Chinese character for 9, '九,' with the index finger. In Japan this gesture means 'thief' so it feels a little strange to us. 10 is represented by crossing the index finger and the middle finger. This forms the Chinese character for 10, '十.' There is also a gesture in which the index fingers of both hands are used to form this cross shape, and the hand may be closed into a 'stone' again to represent 10, or both hands may be opened.

In Taiwan, 6 is the same as in China, but 7 to 9 are different. Also, I thought that it was only in China that they counted up to 10 on one hand, but the Masai tribe in Africa expresses the numbers up to 10 on one hand, and photographs of these hand symbols have been published. The Masai counting method differs from that of China [1], [5].

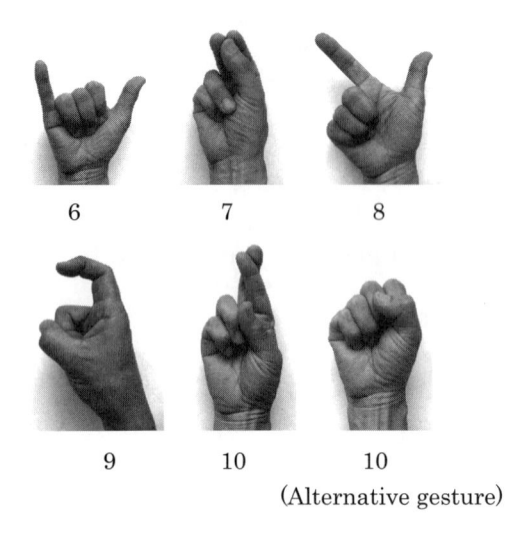

6   7   8

9   10   10
(Alternative gesture)

Figure 3: The Chinese Gestures for Counting 6 to 10

4 The Distribution of Counting Methods on a World Map

When counting with the fingers to communicate numbers to others, we Japanese begin counting from 1 with the index finger held up away from us, but French and German people begin with the thumb. I wanted to know the reason why we ended up different in this way. I was fortunate enough to be granted an opportunity for a year of overseas education in the United Kingdom in 2005. There were foreign students from 40 different countries from around world at Saint Edmund's College in Cambridge. Taking my lunch in the college dining hall, I made conversation regarding counting methods.

It seems that in the German they also start counting from the thumb, and they do so across the whole of Northern Europe. When starting with the thumb, I found that making a 3 was a bit tight. I asked a

German about this and he demonstrated, smoothly opening his fingers, 'Ein, Zwei, Drei.' Then he said that making the Japanese 3 requires some force and is awkward. Hearing this, I noticed that our 3 is indeed difficult to form. We practice from when we are small, and come to think of this movement as perfectly natural. It seems that it is one of those things where it is impossible to say which method is easier.

A Filipino student taking part in this conversation mentioned that they start from the little finger. Since I was thinking that 1 would be made using the thumb or the index finger, this was quite unexpected. In Japan, extending the little finger means 'woman', so this felt a little strange to me. I understood from the literature that there are quite a few countries and regions where counting begins with the little finger [1], [4]. Again, asking someone from Bangladesh, I was shown that the thumb was placed on the first joint of the index finger and moved in order through the joints of the fingers. 1 is made with the finger at the first joint, 2 at the second joint and 3 at the third. Then they counted up to 10 on one hand. This method is also used in India, where the use of the joints of the index finger or the little finger to represent 1 varies by region, but they share the characteristic of using the joints of the finger. When I was in elementary school, we made a game of trying to correctly state the day on which someone was born based purely on their date of birth. At that time, we adopted a method using the finger joints. Perhaps this calculation method had its roots in India or Pakistan.

I have heard an explanation for the reason why Indians can perform this meticulous method of counting, namely, that their thumbs have three joints. Do you really think so? In relation to this, *The History of Numbers* (p22) contains a diagram of a triple-jointed thumb [2]. This concerns a finger arithmetic method illustrated in the 8th century by the English monk, the Venerable Bede. Perhaps there really is a group of people whose thumbs are triple jointed.

As we have seen above, there are many methods for counting, no matter which digit is used for 1. Wouldn't it be wonderful to have a color coding scheme for countries and regions? It occurred to me that with such a graph, one might be able to understand the history of the development of these methods. I was very fortunate to find an excellent reference. Seidenberg performed an investigation in 1960, which reached every corner of the world, and published his findings [4]. He classified every method according to whether it starts on the left little finger, the right little finger, the index finger, the thumb or a finger joint, and drew up a world map. This map is introduced in *Pi in the Sky* [1]. According to the results it might be appropriate to call Africa the origin of humanity, because all of the counting methods exist there. Japan is classified as starting from the thumb in this work. Such a survey requires extensive time and manpower, and if this investigation were performed again, over 40 years later when the internet is prevalent, even more detailed results could no doubt be assembled.

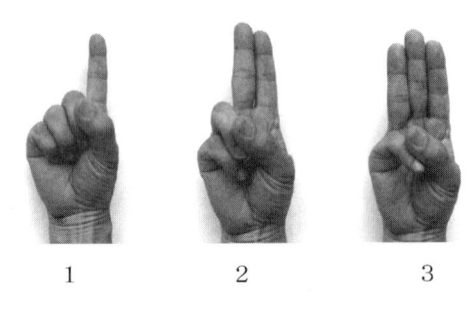

Figure 4: Counting Using the Index Finger (China and Japan)

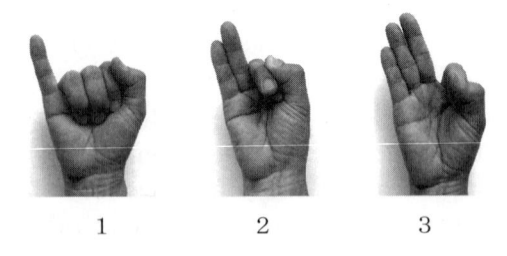

Figure 5: Counting Using the Little Finger (Philippines)

5 In Ancient Rome They Began with the Left Little Finger

We now know that there are differences according to county and region, such as using the little finger, the index finger, the thumb or the finger joints and so on, but this is the way things are today. How did people in ancient times count, and which finger did they start from?

Menniger's *A Cultural History of Numbers* contains the following explanation [3]. Counting on the fingers was for the most part handed

1   2   3

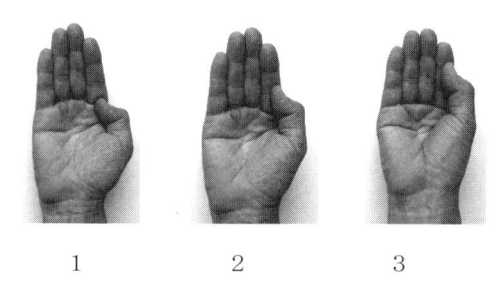

Figure 6: Counting Using the Joints of the Finger (India, Pakistan and Bangladesh)

down by oral tradition. Definitive documents such as textbooks have not been left behind through the time since the Roman era, and the first recorded method was made by the English Benedictine monk, the Venerable Bede. He died in 735 AD, but left behind a document entitled, 'Arithmetic and Conversation using the Fingers.' In this document, 1 is expressed by bending the left little finger into the palm of the hand, 2 is expressed by bending the ring finger down beside it and 3 by bending the middle finger down beside that.

Another counting method using the fingers was illustrated in a document entitled 'The Arithmetic System,' published in Venice in 1494 and written by the Italian mathematician Luca Pacioli. This also expressed 1 by folding down the left little finger. This indicates that from ancient times up until the Middle Ages, when people counted on their hands, they began in general with the left little finger. This fact seems particularly unusual to me, as I was thinking that counting from 1 would begin with the thumb or the index finger, and that even if it began with the little finger, I thought it would involve *opening* the fingers. I wonder why they used the left hand? Also, why did they go from the little finger? Also, another question emerges. I found the method which involves bending the left little finger hard to perform. In West Africa, bending the little finger on the left hand is indeed a 1, but there is a diagram in Zaslavsky's book of illustrations which explains pictorially that in this case the index finger of the right hand is used to push down the left middle finger [5]. Perhaps it is difficult to bend it by itself after all? Or perhaps people in ancient times learned to bend it through practice.

I have previously discussed the fact that computers are based on binary. I have heard it explained that if we used this binary system with both hands, we could count to 1024. If an extended finger represents 0 (zero) and a bent finger represents 1 (one), then each finger represents 1 bit, and since we have 10 fingers we can express a 10-bit number. Since $2^{10} = 1024$, it was explained that we could count up to 1024, but this is truly the kind of topic you'd expect a computer scientist to enjoy, and it's doubtful whether we could really manage this kind of finger bending. The Venerable Bede and Luca Pacioli both explained and illustrated how to count as far as 10,000 using both hands. This is well beyond 1024. In addition, they didn't just explain how to count, but also how to perform multiplications and so on with the fingers, so we can say that this was indeed a technically well-developed field.

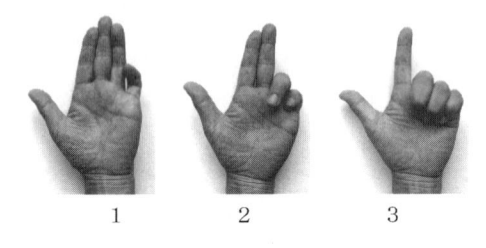

Figure 7: Counting Using the Little Finger of the Left Hand (Ancient Rome)

6 Food Culture, Lifestyle and Counting

Searching the internet, it was written that across the world there are 27 types of counting method using the fingers. Wouldn't there be even more considering the possible combinations? Counting using the fingers differs according to region, ethnicity and historical period. But why did the methods change in these ways and eventually arrive at the present situation? It's fun to think about this 'evolutionary theory' of counting methods from various angles.

To begin with, the ancestors of humanity, the Africans, possess all of the counting methods. Expressing 1 with the thumb, with the index finger, the right little finger, the left little finger, *etc.*, all of the archetypes exist there. There are also many tribes such as the Masai who count from 1 to 10 using one hand, and counting systems can be said to be considerably advanced. Also, of particular note, not only base

10, but also counting methods and numerical notation systems using the principle of binary exist there. Do these complex ways of presenting the fingers according to binary patterns constitute evidence of an ability to move the fingers freely? There are also tribes that have adopted base 20. Perhaps this owes to the fact that a barefoot lifestyle makes the toes available for use too. This topic is dealt with in more detail by Zaslavsky [5].

We Japanese do not use binary in the same way as the African peoples. In that regard, counting methods may be considered to have developed so that we don't only rely on the fingers, but use other methods such as calculators, *etc.* It's also possible to imagine that the reason we don't use base 20 is because we don't have bare feet. By starting to wear shoes or other such footwear that hides the toes away, we settled on base 10, which uses the hands alone.

In ancient Rome and in the Middle Ages, bending the left little finger was 1. Maybe the reason for using the left hand was that the right hand was used for more important things. Also, bending the little finger is difficult, and yet we don't know the reason for its use. As time advanced, people began to communicate 1 using the index finger in China and Japan, using the thumb in Germany and France, and using the little finger in the Philippines, *etc.* Counting methods are a part of our national and ethnic identity. Language, religion, currency, systems of measurement and so on are characteristics of our ethnicity. Just as other differences express differentiation among peoples, perhaps a conscious awareness of these differences led to the differentiation of counting methods.

Maybe the difference between starting from the open 'paper' condition and proceeding by bending the fingers, versus starting from the closed 'stone' condition and extending the fingers, is related to the air temperature. Perhaps the natural position of the hand is either 'paper' or 'stone' for different races. In hot countries it would be 'paper' and in cold countries it would be 'stone.' According to such thinking, Africa, being a hot country, would count from 'paper,' while the tribes who moved to colder Northern countries would count from 'stone.'

While conversing about the different counting methods with the foreign exchange students in the college dining hall, the following thought suddenly crossed my mind. Perhaps the ways of counting with the fingers are related to food culture. Meals in Europe involve a knife and fork. When cutting meat it is necessary to firmly grasp the knife and fork. When so doing, it is the thumb which is free to move, so is this why extending the thumb came to represent 1? Also, the countries in

Northern Europe are cold, so many people keep their hands closed and perhaps the easily extended thumb naturally came to represent 1.

Once a method of counting with the fingers is established, the other fingers begin to take on other meanings. In the West, the thumb represents 1, but in Japan it indicates 'man.' In the Philippines, the little finger represents 1, but in Japan it indicates 'woman.' The Chinese 9 indicates 'thief' in Japan. On the other hand, the Japanese 2 is the 'victory' sign in the United Kingdom. It is common for the meanings of the fingers to be completely different in different countries. The English zoologist Desmond Morris researches this in his book *Bodytalk: A World Guide to Gestures*. Interested readers should refer to the references. Counting methods using the fingers are not limited to the expression of numbers, they appear to go as far as influencing each county's culture.

References

[1] J.D. Barrow, (trans. M. Hayashi) *Tenku no Pi [Pi in the Sky: Counting, Thinking, and Being]*, Tokyo: Misuzu Shobo, (2003).

[2] D. Guedj, (trans. I. Nanjyo), *Su no Rekishi [L'empire des nombres]*, Osaka: Sogensha, (1998).

[3] K. Menninger, (trans. M. Uchibayashi), *Zusetsu Su no Bunkashi [Number Words and Number Symbols: A Cultural History of Numbers]*, Tokyo: Yasaka Shobo, (2001).

[4] A. Seidenberg, *The Diffusion of Counting Practices*, Univ. of California Press, (1960).

[5] C. Zaslavsky, *Africa Counts: Number and Pattern in African Culture*, Lawrence Hill Books, (1999).